CLASSROOM ASSESSMENT IN MATHEMATICS

Views from a National Science Foundation Working Conference

Edited by

George W. Bright
Jeane M. Joyner

This project was supported, in part,
by the
National Science Foundation
Opinions expressed are those of the authors
and not necessarily those of the Foundation

University Press of America,® Inc.
Lanham • New York • Oxford

Copyright © 1998
University Press of America,® Inc.
4720 Boston Way
Lanham, Maryland 20706

12 Hid's Copse Rd.
Cummor Hill, Oxford OX2 9JJ

The work reported here was supported by the National Science Foundation (Grant number ESI-9618531). All opinions and positions taken however, represent the views of the individual authors and do not necessarily reflect the positions of the Foundation or any other government agency.

Library of Congress Cataloging-in-Publication Data

Classroom assessment in mathematics : views from a National Science Foundation working conference / edited by George W. Bright, Jeane M. Joyner.
p.   cm.
Includes bibliographical references.
l. Mathematics—Study and teaching—Evaluation—Congresses. I. Bright, George W. II. Joyner, Jeane M.
QA11.A1C57 1998 510'.71 —DC21 97-51761 CIP

ISBN 0-7618-1027-7 (cloth: alk. ppr.)
ISBN 0-7618-1028-5 (pbk: alk. ppr.)

Classroom assessment in
mathematics

Table of Contents

Introduction from National Science Foundation

Assessment is the means by which we determine what students know and are able to do. Sometimes it seems that the amount of attention given to student assessment in our society is directly proportional to the stakes associated with the assessment. High stakes assessment, therefore, garners much attention. Yet, these "high stakes" tests are episodic in nature and frequently administered toward the end of a school year, so there is relatively little opportunity to use test results to improve student thinking and performance in areas of weakness. The type of reporting that often accompanies test results compounds the problem further: many tests do not give reporting levels that inform individual teachers about the strengths and weaknesses of their individual students. All of this accentuates the need for high quality classroom assessment that is both informal and formal.

Informal classroom assessment allows a mathematics teacher to ascertain whether students have a sufficient understanding of a concept and readiness to extend it further or whether more work -- and perhaps even what kind of work -- is needed to first "cement" basic understanding and performance. Formal classroom assessment provides needed evidence of a more summative nature on behalf of what each student understands and can do at a particular point in time. Teachers need to understand and experience these "best practices" in classroom assessment that are informed by research. While some best practices are applicable in a variety of disciplinary contexts, others are quite specific to mathematics.

The conference, *"Classroom Assessment in Mathematics,"* was designed with a focus on quality day-to-day classroom assessment in mathematics, in order to help teachers (a) learn frameworks to make appropriate instructional decisions based on real knowledge of students' thinking, (b) develop effective assessment skills to gather and organize quality evidence of students' understandings, and (c) learn effective strategies to inform students and their families about content goals and students' progress toward attaining those goals. The National Science

Foundation is pleased to have provided funding for the conference We hope and expect that its work will inform and challenge the mathematical and educational communities about not only the growing importance of assessment at the classroom level but also its evolving knowledge base.

Margaret B. Cozzens
Director
Division of Elementary, Secondary, and Informal Education
National Science Foundation

Introduction from North Carolina Department of Public Instruction

The importance of teachers being able to assess where their students are cannot be emphasized too strongly. Teachers must understand the strengths and areas of need of their students in order to help students achieve.

Knowing what students understand and what they are able to do allows teachers to adapt their instruction so that curricular goals are met with a high level of proficiency by students in their classes. To provide the greatest opportunities for learning, teachers must have a sense of how instruction can be adapted for their entire class, for groups of students, and for individuals. Classroom assessment provides teachers with a process for gathering this information. It involves gathering and utilizing diagnostic information about individual students, obtaining feedback about how individuals and groups of students are progressing toward learning goals, and gathering and interpreting data for making summative evaluations.

On-going assessment has always been part of strong instructional programs. It is at the heart of the teaching-learning experiences that take place daily in classrooms everywhere. It is the at the core of teacher-student interactions. However, classroom assessment as it informs instructional planning and monitors students' progress is also in danger of being lost in the movement for high stakes accountability. Teachers sometimes abandon classroom assessment as a way to gain insight into students' thinking and reasoning in favor of "practicing for the test." These decisions may be made because of perceptions that there is no longer time to engage in classroom assessment. But to make instruction more effective and to help more students achieve at higher levels, it is essential that teachers continue to use classroom assessment as an important way to monitor the learning of students.

This conference, sponsored the National Science Foundation, the North Carolina Department of Public Instruction, The University of North Carolina at Greensboro, the University of North Carolina

Mathematics and Science Education Network, and the Greensboro Area Mathematics and Science Education Center, focused on the importance of classroom assessment as part of the bigger picture of educating students. The papers of the participants and discussions during the conference can be instrumental in providing a knowledge base so that teachers, researchers, and policy makers will be better informed about the potential of classroom assessment in strengthening instructional programs. Through the recommendations, leaders in every aspect of education have been given suggestions for actions they can take to support teachers in developing and expanding their understanding and use of classroom assessment.

It was a pleasure for the North Carolina Department of Public Instruction to be a co-sponsor of this important conference. On behalf of the Department, I hope that the work represented in this proceedings volume will stimulate both discussion and action, so that classroom assessment continues to be valued as an important part of teaching.

Henry L. Johnson
Associate Superintendent
North Carolina Department of Public Instruction

Introduction from Co-Principal Investigators

Planning and implementing this conference was a professionally rewarding experience. We have learned a great deal about classroom assessment, and we had opportunities to interact with many of the best thinkers, from the perspectives of both research and practice, in this emerging field.

There are many groups and people that we need to thank for supporting us in this effort. We are grateful to the National Science Foundation (NSF) for funding our project. Without their financial support the conference would never have taken place. Specifically, we thank Diane Spresser and Jean Vanski of NSF for participating in the conference and being so encouraging about our work.

We are grateful to the North Carolina Department of Public Instruction (NCDPI) for providing support in a variety of ways: contributed time of NCDPI personnel, meeting space, vans which were our transportation between the hotel and the meeting site, and many other items.

We are grateful to The University of North Carolina Mathematics and Science Education Network, the Greensboro Area Mathematics and Science Education Center, and The University of North Carolina at Greensboro (UNCG) for moral as well as logistic support. Their help was essential to our work in carrying out this project.

We want to acknowledge individually some of the people who provided the on-site support that is so essential for a successful conference. Specifically, we want to thank Mike Kestner, Toni Meyer, Paula Langill, Rick Klein, and David Mills from NCDPI for their help both before and during the conference. We thank Melanie Nickerson at UNCG for her help with mailings, preparation of conference materials, word processing, and overall encouragement.

Most of all, though, we want to thank the participants for taking time from their busy schedules to spend almost three days in North Carolina thinking and talking about classroom assessment. Their input was obviously essential to the success of our discussions. We thank them for drafting working papers and then returning home and revising their thoughts in the form of the working papers which are archived in this volume. It is important to have this documentation of their thinking for others to access.

We hope that this volume sparks more discussion about classroom assessment that leads to action to improve classroom assessment in every mathematics classroom in the country. After all, it is the learning of students that we ultimately want to improve.

George W. Bright
Jeane M. Joyner
Fall 1997

Section 1

The Conference

Design of the Conference

George W. Bright
The University of North Carolina at Greensboro

Jeane M. Joyner
North Carolina Department of Public Instruction

In 1997, the National Science Foundation (NSF) funded a proposal (Grant No. ESI-9618531) whose purpose was to identify research issues and implementation strategies which support quality classroom assessment. Significant, in-kind support was provided by the North Carolina Department of Public Instruction (NCDPI). Further contributions were made by The University of North Carolina at Greensboro (UNCG) and the Greensboro Area Mathematics and Science Education Center (GAMSEC), which is part of The University of North Carolina Mathematics and Science Education Network (MSEN). The strategy taken by this project to fulfill its purpose was planning and conducting a conference whose participants were asked to explore the potential richness of classroom assessment in order to help teachers (a) learn frameworks to make appropriate instructional decisions based on real knowledge of students' thinking, (b) develop effective assessment skills to gather and organize quality evidence of students' understandings, and (c) learn effective strategies to inform students and their families about content goals (feedforward) and students' progress toward attaining those goals (feedback).

If the educational system is to help more students achieve at higher levels in mathematics, greater attention must be focused on day-to-day linkages among teaching, learning, and assessing. Student performance is influenced by instruction, and one way that instruction can be improved is by helping teachers develop better information about students' thinking. Day-to-day classroom assessment provides a vehicle for teachers to use to acquire knowledge about student performance and thinking. With high quality classroom assessments, teachers can help students develop mathematical power that will be revealed in greater student achievement.

In the past decade, there has been a great deal of attention paid to ways of obtaining information about students' thinking and performance. Reform documents, for example *Professional Teaching Standards* (National Council of Teachers of Mathematics, 1991), have focused on "discourse" as a critical factor in enhancing student understanding and for helping teachers learn about their students. The use of questioning, with the accompanying need for wait time, is one strategy among others (e.g., Hiebert & Carpenter, 1992) that teachers can use to gain information both about students' thinking and about how students' performance increases when that information is used by teachers to improve instruction. Recent recommendations about the use of alternative assessments have been driven in large part by the desire for more students to display the depth of their knowledge in multiple settings. Allowing teachers to see the range of application of students' knowledge is important if teachers are to obtain a complete picture of what students know and can do. More effective instruction should be built on as complete a picture of students as is possible to obtain.

High quality classroom assessments, partly because they are continuous and responsive to the changing needs of students, are likely to have great impact on the ultimate learning of mathematics. If students are given clear learning targets (feedforward) and helpful, timely feedback about the quality of their work, and if as a result they develop solid understanding of the content, they are less likely to forget what they have learned and they are more likely to be able to use what they have learned in new and challenging situations. Conversely, the lack of such quality classroom assessments and the likelihood of instruction that fails to respond to students' emerging understandings may negatively impact the ultimate achievement of each learner and have powerful long-term negative consequences.

Along with providing information for students, high quality classroom assessment should also inform the teacher and students' families. It seems reasonable to expect that there will be stronger, on-going encouragement from home for students to put forth the needed effort to master mathematical ideas and to remain engaged in mathematics course-taking if students know what the learning targets are and have an understanding of what quality performance looks like, if teachers better understand the strengths of their students, and if teachers can share this information with students' families.

Background

Increasingly, research on students' knowledge of mathematics suggests that many students do not understand mathematics in the same ways that their teachers do. This means that there is an increasing burden on teachers to elicit explanations from students about their thinking; otherwise, teachers cannot know what their students actually understand. Likewise, in order for teachers to incorporate this knowledge about students into their instruction, they must begin to help students build mathematical ideas from their own thinking. Teachers must develop skill in developing or choosing assessments that elicit students' thinking and then interpreting what students say in light of the teacher's more sophisticated knowledge of mathematics. Fortunately, there are some recent examples of how these strategies can be implemented. These examples can help guide teachers in beginning the process of understanding their students' thinking and then planning instruction that supports the development of more sophisticated thinking.

At the elementary school level, Carpenter and Fennema have organized this process in an inservice program called Cognitively Guided Instruction (CGI). They have shown that teachers can acquire very specific information about specific student's mathematical thinking and then can use that information to set tasks for individual students (and groups of students) that encourage use of increasingly sophisticated strategies for solving problems. Implementation of CGI also results in higher performance on both standardized tests and problem solving tests (Carpenter, Fennema, Peterson, Chiang, & Loef, 1989; Fennema, Carpenter, Franke, Levi, Jacobs, & Empson, 1996; Fennema, Carpenter, & Peterson, 1989; Fennema, Franke, Carpenter, & Carey, 1993; Peterson, Fennema, Carpenter, & Loef, 1989).

Cobb and others (e.g., Nicholls, Cobb, Yackel, Wood, & Wheatley, 1990) have also developed techniques that focus on helping teachers understand students' thinking in order to adapt instruction for the students at hand. Students in these programs perform better on tests; but perhaps more critically, students develop better dispositions toward doing mathematics and engaging in mathematics tasks. The reports of this project have noted that in order to be successful, teachers need to engage in continuous classroom assessment of students' understanding.

> For a teacher who seeks to foster students' higher order
> mathematical thinking, no conventional achievement test
> is likely to be of immediate help. *What teachers can use*
> for this enterprise is more-or-less constant feedback on
> how students interpret the problems before them.
> (Nicholls, et al., 1990, pp. 137-138; emphasis added)

At the middle school level, one notable recent attempt to help
teachers learn to listen to students' thinking is the QUASAR project
(e.g., Silver & Smith, 1996). Again, student performance improved,
and teachers began to develop understanding of differences between their
thinking and their students' thinking. Silver and Smith note, however,
that there are many ways for teachers to assume the trappings of
techniques for understanding students' thinking without really helping
students develop sophisticated mathematical reasoning.

> Although there were important mathematical issues that
> could have been addressed [in the class discussion] ...
> these issues were virtually ignored in the students'
> questions and comments.... As students' confidence and
> comfort grows, the teacher must ensure that the
> mathematics does not get lost in the talk and that progress
> is made along the path not only toward a real
> mathematical discourse community but also toward the
> increased mathematical proficiency of all students. (pp.
> 23-24)

That is, there is no guarantee that focusing on students' thinking alone
will result in better instruction that promotes better learning of
mathematics.

The Assessment Communities of Teachers project (Briars, 1996)
focuses on different ways school districts provide support for teachers as
they learn to use alternative methods in assessing their students. While
the project does not yet have definitive findings, the underlying
framework of the project is that assessment can be a vehicle for
furthering mathematics reform and that by modifying assessment
practices instructional practices will be positively impacted.
Alternative assessment methods are seen as a way to better understand
not only how students are thinking and reasoning but also what sense
they are making of mathematics.

There is less evidence at the high school level that instructional decisions can be modified based on teachers' knowledge of students' thinking. The evaluation of the Interactive Mathematics Project (Webb, 1996), currently in progress, may provide insight into this issue. However, the embracing of the "Chicago materials" (McConnell, Brown, Eddins, Hackworth, & Usiskin, 1990) and the Hawaii algebra materials (Rachlin, Matsumoto, & Wada, 1992) suggests that at least some high school teachers are searching for ways to engage students in mathematical tasks that do build on global knowledge about students' thinking. It seems likely that teachers who gain increased understanding of students' thinking will make instructional decisions that produce better performance and understanding by those students.

Need for Improving Classroom Assessment

Our assumption is that teachers must come to understand how students create mathematical knowledge and to use that understanding in making decisions about appropriate instruction. Romberg, Zarinnia, and Collis (1990) seemed to support this view when they noted that "because the intent [of a new model of assessment] is to assess the creation of knowledge and the processes involved rather than to measure the extent to which students have acquired a coverage of the field of mathematics, a much wider variety of measures, many of them qualitative, are needed" (p. 35). That is, teachers need access to information about students' thinking that can be revealed only through improving the types of classroom assessments currently in use.

When students create knowledge, they must do more than merely internalize concepts and procedures. At a minimum, creating mathematical knowledge involves the development of higher order thinking. Resnick (1987) describes higher order thinking by noting that it is nonalgorithmic, tends to be complex, often yields multiple solutions, involves nuanced judgment and interpretation, involves application of multiple criteria, often involves uncertainty, involves self-regulation of the thinking process, involves imposing meaning, and is effortful. Many of these characteristics are best observed in day-to-day mathematical activity, so the classroom teacher is clearly in the best position to monitor the development of higher order thinking. One vehicle for monitoring is continuous classroom assessment with tasks that the teacher knows are appropriate for the students, rather than through summative testing at the end of a school year using tasks that are set by someone unfamiliar with the students.

If mathematics reform is to succeed, teachers must know about and use a wide range of assessment tasks, so that they can come to understand the complexity of students' thinking about mathematics. The conference conducted in the project reported here was designed to articulate issues and conerns related to helping teachers understand and learn to use these new types of assessments.

Existing Understanding of Classroom Assessment

Assessment skills. While insightful teachers have long employed informally a variety of performance assessments (i.e., projects, debates, investigations, and student work folders), utilization of these strategies has been uneven at best. Teachers' competence to perform a variety of quality assessments and to make appropriate inferences lies at the heart of high quality classroom assessment. "To a much greater degree than in traditional assessment, the quality of alternative assessments will be directly affected by how well teachers are prepared in relevant assessment skills" (Worthen, 1993, p. 448). In addition to understanding mathematics, teachers must become expert implemeters of classroom assessment.

One of the important issues for teacher enhancement is to know how teachers learn to conduct high quality classroom assessments. There may be a sequence of "development" (or at least stages of growth) that teachers generally go through as they become expert at managing classroom assessment; for example, (a) using tasks that are intended only as assessment tasks followed by (b) using tasks that are designed to integrate assessment with teaching. How can we design sample tasks? How can teachers design sample tasks? What are some characteristics of each type of task? The work of the Balanced Assessment project (as outlined by Sandra K. Wilcox at the RAND/NSF Conference on Performance Assessment in Science and Mathematics in late 1995) and the New Standards project (as outlined by Phil Daro at the same conference) seem likely to inform this issue.

But even if a teacher accepts the notion that students' views are important to understand, the way that a teacher interprets the evidence of students' thinking is greatly influenced by the teacher's assumptions about mathematics, learning, learners, and what constitutes convincing evidence.

> More often than not ... [teachers'] assumptions, beliefs, and values are hidden from consciousness and limit the capacity to be systematic. A key to being systematic

> about integrating assessment into instruction is knowing how to manage one's personal stamp, starting with a planned process of selecting assessment activities, a process that also tests one's assumptions and reflects one's values and beliefs. (Driscoll, 1995, p. 421)

Teachers must constantly compare what students say with what those teachers think students know. That is, teachers must infer what students know from what students say and do. But each teacher has to remember that these inferences are based on one's "personal stamp." Each teacher's knowledge and beliefs act as a filter which restricts what the teacher thinks students are saying and doing.

Assessments that are used as indicators of achievement (e.g., SATs, state tests) are often fairly global. How does maintaining a focus on continuous classroom assessment change a teacher's view of efficacy about teaching? That is, what are the mileposts that we can suggest to teachers as evidence that they are obtaining useful information, that they are interpreting the information effectively, and that their instructional decisions are responsive to students' needs? It is clearly not sufficient to suggest that teachers should wait until they receive scores back on such global measures of performance. If types of high quality classroom assessment can be identified clearly, if techniques for helping teachers learn how to employ these assessments in their instructional decision-making and in the feedback they give to students can be outlined, and if the public can be convinced that quality classroom assessment takes time to implement but should be highly valued for its contribution to the ultimate success of students, then it seems reasonable to expect that over the long term, student performance might be greatly improved.

If there is any downside to the implementation of high quality classroom assessment, it is that teachers need time and experience in clearly articulating goals, designing quality assessments, and learning how to take advantage of the information they acquire about students' thinking (e.g., Fennema, Carpenter, Franke, Levi, Jacobs, & Empson, 1996). The more teachers become immersed in considering students' thinking processes and their progress toward mathematical goals as well as the ultimate products of their work, the more the complexities of designing and using wisely classroom assessment become apparent. Teachers need several years and opportunities to work collectively to take advantage of their increased knowledge of students and assessment.

Technology and assessment. Technology has the potential both to complicate and to enhance classroom instruction and assessment. Computers currently provide opportunities for preprogrammed, individualized assessment. Many computer-assisted assessment packages provide sequenced content and varying levels of difficulty; student progress is automatically tracked by the programs and, in many instances, student error patterns are analyzed to determine needed instructional review and reteaching. How skilled teachers are in choosing and utilizing technology to facilitate higher order thinking (as opposed to using it only for routine drill and practice) and in taking advantage of the individualized feedback and record-keeping may become equity issues. These issues may further separate classrooms in which teachers use high quality assessment to make instructional decisions and give students helpful feedback from classrooms in which technology is poorly utilized or non-existent.

Calculators seem to change the types of decisions students make about how to solve problems (Hembree & Dessart, 1986, 1992). For example, when calculators are present, students are more likely to choose the correct operation for solving problems, even though they are not required to solve the problem. Because calculators are more readily available than computers, there is an immediate need to help teachers address assessment issues (e.g., problem solving, equity) in situations where students have access to calculators.

In short, technology changes both the amount and the nature of interactions among students and between students and teachers. It is reasonable to expect that techniques of continuous classroom assessment might be different if technology is present and well utilized.

Feedforward and feedback enhance learning. One of the interpretations that might be made about our current understanding of classroom assessment is that when teachers try to understand students' thinking and then adapt instruction to build on that thinking, students also learn more about their own thinking. That is, when students and teachers talk about goals and what constitutes quality performance, students better understand the learning targets. When they hear how they and their peers solve problems, they have opportunities to expand their own thinking.

An important technique that teachers use to provide feedback to students about their thinking is to create opportunities for students to evaluate their own solutions. Cobb and his colleagues noted that "when, as frequently happened, students offered two or more conflicting solutions, the teacher typically framed the situation as a problem for the students to resolve by justifying and explaining their solutions" (Nicholls, et al., 1990, p. 141). By highlighting the conflict among solutions (as opposed to the conflict among students), teachers can implicitly tell students that the solutions are different, that the solutions are contradictory, and that students need to assume responsibility for determining which solution, if any, is appropriate. This last point is also supported by Driscoll (1995), who noted that giving feedback allows students to take control of determining if their thinking is applicable across problem situations, rather than leaving this control in the hands of the teacher.

A common misconception of many teachers of mathematics is the assumption that students' knowledge is like the teacher's. In this view, teachers can adopt a "teaching as telling" vision of teaching (e.g., Smith, 1996) in which there is no need for continuous classroom assessment. Students are viewed more as members of a group, likely to think and reason in ways that mirror the teacher's guidance, rather than individuals who bring their past experiences and knowledge and unique ways of thinking to the mathematical lessons being taught. If this is the view that teachers adopt, there is little need to know precisely what students are thinking, since students' views are interpreted in terms of the teacher's views.

The alternate view that we take is that knowledge is actively built in a dynamic process based on effective communication between teachers and students. Classroom assessment must capture this process in a continuous, interactive way, and teachers must work very hard to make sense of what students are doing. *"Knowledge of the mathematics of students is not passively received, but is actively built up by teachers"* (Steffe & D'Ambrosio, 1996, p. 65, emphasis in the original). Teachers must provide tasks that will elicit thinking and then listen carefully to the ways that students explain how they dealt with those tasks. There must be interaction between the teacher and the students so that each can come to understand what students are thinking; this is precisely the notion of continuous classroom assessment.

Integrating continuous classroom assessment with instruction is surely not easy. Indeed, "merely" teaching with cognitively complex tasks is itself a cognitively complex task.

> To be successful in implementing cognitively demanding tasks in the intended manner, teachers may need to use a combination of pedagogical strategies, including modeling high-level performance, supporting students by performing part of the task and leaving the remainder for them to do, giving students sufficient time to generate and explore their own ideas, encouraging student self-monitoring, and consistently communicating the need for explanation, meaning, and understanding. (Silver & Smith, 1996, p. 26)

Further, we need to consider how we can help principals and families begin to value classroom assessment because it helps students. One way to help them value classroom assessment is to demonstrate that instruction improves when teachers have evidence which conveys what students know and can do. "Progress in mathematics teaching, learning, and assessment must take place in concert with the goals and standards that are valued by the mathematics community as well as by the public, through a shared process of meaningful study and dialogue" (Kulm, 1990, p. 77).

Students as active partners. Many teachers will need to rethink the ways they conceptualize their work, including (but certainly not limited to) the ways that they interact with students. Making students more active partners in the teaching-learning process through feedforward and feedback, developing skill in using a variety of assessment strategies to better understand students' thinking and to determine the mathematics that students are able to apply, and making decisions about instruction that facilitate learning for individuals within the context of the total class are assessment challenges which we believe require support and professional development for teachers.

The Bronx New School is one example of an attempt to put teachers at the heart of assessment of students' thinking.

> The assessment system at the Bronx New School (BNS) emerged out of a conception of teaching that places students at the center of the learning environment. Classrooms are structured to encourage active inquiry and are stocked with a wide range of concrete materials meant to be used for direct investigation.... Students in BNS classrooms are also regularly engaged in opportunities to exchange ideas and to collaborate with peers.... To put

this vision of education into practice, teachers need to understand human development and learning theory, content matter and teaching strategies, and, most of all, their students. Much of this information derives from an assessment approach that places observation of students and their work at its center. As teachers observe what their students know and can do, the particular strategies their students use as they learn new things, they then use this information to build bridges between past and future skills and understandings. (Darling-Hammond, Ancess, & Falk, 1995, pp. 206-207)

Implementation of this approach on a wide-scale basis would seem to require considerable change in the positions of the public and of politicians in allowing teachers to take responsibility for learning what students know.

Conference Details

Purposes and Audiences
The purpose of this project was to identify research issues and implementation strategies which support quality classroom assessment as a tool for greater student achievement. The recommendations generated by the conference are of interest to three main audiences:

- researchers and teacher enhancement project directors
- preservice and inservice teacher educators
- personnel in state departments of education

Researchers and teacher enhancement project directors will be better informed about issues which need to be dealt with in teacher enhancement proposals. Institutions of higher education, school systems, and staff development projects developing preservice and inservice programs with classroom assessment components will be better informed of issues which need to be included in their plans. State departments of education staffs, especially those engaged in large-scale accountability programs, will be better informed about the need to include a component that focuses on classroom assessment as part of any comprehensive assessment system.

State departments of education have a particular important use of the information generated by the conference. Many state departments are charged with responsibility for monitoring student performance in mathematics through some form of statewide assessment program. We believe that a statewide assessment system would be more comprehensive, effective, and helpful to teachers if it included a component that emphasizes high quality, day-to-day classroom assessment along with the more traditional summative evaluations. Such a component would help inform everyone (e.g., teachers, students, families) about the continuous development of student understanding so that adjustments could be made as conditions required, rather than in response to evidence of some "crisis" in learning that is identified only after results from a standardized test are made public. This conference can help state departments examine and suggest classroom assessment practices that are likely to support student achievement and greater success on summative (i.e., statewide) assessments.

Conference Attendees and Framing Questions

Conference attendees (see Section 4 of this book) were invited because of their expertise in areas relevant to the conference objectives; namely, expertise related to exploring classroom assessment in order to help teachers (a) learn frameworks to make appropriate instructional decisions based on real knowledge of students' thinking, (b) develop effective assessment skills to gather and organize quality evidence of students' understandings, and (c) learn effective strategies to inform students and their families about content goals (feedforward) and the students' progress toward attaining these goals (feedback). Participants were chosen from research, assessment, and practicing communities so that different aspects of classroom assessment would be articulated. Part of the charge to participants was that discussions should clarify primary and secondary issues surrounding the creation and wise use of classroom assessments as effective links among teaching, learning, and assessing.

A variety of issues were identified that were expected to surface during conference discussions. The agenda was organized to facilitate discussion around several of these issues.

1. The need for teachers to articulate clear short-term and long-term learning goals (and how teachers make these decisions).
2. The importance and nature of quality feedback.
3. The use of both alternative and traditional assessment methods to gather better information about students' learning (the skills teachers need and the balance of assessment methods).
4. Teachers' wise use of assessment information to improve instruction and the professional development needs for this to

take place (gathering and organizing evidence and making sound decisions based on the evidence).

5. Ways to use technology in assessing students' understandings.
6. Ways to engage students in setting personal learning goals, exploring the depth of their own understanding, and evaluating and documenting their own progress.
7. Ways of informing families about the development of students' knowledge of mathematics.
8. Ways of demonstrating the value and usefulness of quality classroom assessment to the public as an important component of documenting student achievement.

As a means of initiating discussion, each participant was invited to write a working paper, to be distributed prior to the start of the conference. Three participants (Diane Briars, Mari Muri, and Norman Webb) were asked to write position papers that would address, respectively, three general areas: school district implementation issues, state department of education issues, and general theoretical frameworks. All other participants were asked to address some or all of the questions given in Figure 1. After the conference, participants were asked to revise their papers. The final versions are included in this monograph.

Based on your experiences as a teacher, working with teachers, and your knowledge of day-to-day classroom assessment, please respond to the following questions:

- What kind of information do teachers need to provide students about the learning targets? How do teachers establish expectations and goals? What is the role of students?
- What kinds of information about students do teachers need to gather in their daily interactions with students? What constitutes quality evidence?
- What techniques seem most useful in helping teachers gather that information and what skills do teachers need in order to implement these strategies?
- How can teachers learn to organize and interpret that evidence appropriately in order to make sound judgments about student learning and achievement?
- What is known about helping teachers learn to make instructional decisions based on that evidence?
- What constitutes quality feedback and in what ways can teachers provide useful feedback to students and their families?

Figure 1. Framing Questions Given to Participants

Organization of Agenda

The conference was held on May 16-18, 1997. Formal sessions were held on Friday afternoon, Saturday morning and afternoon, and Sunday morning. A banquet was held on Friday evening; after the meal, a student, a parent, a school administrator, and a representative of NSF were asked to present their perspectives on classroom assessment. Comments by the student and the NSF representative are reproduced in Appendix A and Appendix B, respectively. The conference concluded on Sunday with an opportunity for all participants to synthesize the discussions and to make recommendations.

The conference agenda (Figure 2) was developed after the working papers had been received and distributed to participants. Early in the conference, authors of the position papers were given a brief opportunity to expand on their ideas and to react to the working papers. Then discussion focused around categories of issues that seemed critical for understanding classroom assessment.

	Friday, May 16
1:00	Welcome and Introductions; Review conference charge
1:30	Comments by Special Guests
1:45	Comments by Position Paper Authors
2:30	Working Groups ONE: Background Issues Related to Classroom Assessment
4:30	Reporting Back on Working Groups ONE
	Saturday, May 17
9:00	Working Groups TWO: Gathering and Interpreting Evidence
10:45	Working Groups THREE: Instructional Decision Making
12:00	Lunch
12:45	Reporting Back on Working Groups TWO and THREE
2:00	Working Groups FOUR: Communicating About Classroom Assessment
3:45	Reporting Back on Working Groups FOUR
4:30	Open Discussion on Other Issues
	Sunday, May 18
8:45	Reactions and Identification of Major Issues
9:30	Open Discussion
10:30	Working Assignment: In small groups, summarize the most critical issues and describe what you think is most important to communicate about those issues.
11:15	Brief Reporting Back from Groups
11:40	Wrap Up

Figure 2. Agenda

Questions (Figure 3) were prepared for each Working Groups session; these questions were intended to help initiate discussion. In addition, at each working group session participants were charged with recommending some "starting points" for addressing the issues raised at the conference. During the conference it became clear that it was vital to create a common understanding of "classroom assessment." This discussion became the focal point of Working Groups Two and Three.

Working Groups ONE: Background Issues Related to Classroom Assessment
- What information do teachers need in order to inform their instruction?
- How do teachers' knowledge of and beliefs about mathematics content and curriculum impact classroom assessment practices?
- How do teachers' knowledge of and beliefs about learning impact classroom assessment practices?
- How does involving students in goal setting and evaluating their achievement of goals impact classroom assessment practices?

Working Groups TWO: Gathering and Interpreting Evidence
- How can teachers use a variety of techniques to reveal different aspects of students' thinking and ability to use mathematics? What is the role of technology?
- How do teachers become better at choosing or creating good questions and good tasks? How are those choices influenced by technology?
- How can teachers record and organize assessment information? What is the influence of technology on this?
- What factors influence the inferences that teachers make from the evidence?

Working Groups THREE: Instructional Decision Making
- How can teachers become better at connecting inferences about children's performance and thinking to instructional decision making?
- How can linking assessment and instruction address equity issues related to student differences?
- How can teachers learn to deal with the "mixed messages" that different kinds of assessments might provide?
- Is use of assessment information different when instruction is traditionally versus technologically based?

Figure 3. Questions Posed to Working Groups (continued on next page)

Working Groups FOUR: Communicating About Classroom Assessment
- How can students become more involved in the assessment process (e.g., clarifying learning targets, defining quality performance, evaluating their level of performance)?
- How can teachers better communicate expectations of assessment and give quality feedback to students? How can feedback serve instructional purposes?
- How can we communicate the importance of classroom assessment to teachers, administrators, policy makers, families, etc.?
- How does connecting classroom assessment with instructional decision making support improved achievement among all students?

Figure 3.   Questions Posed to Working Groups (concluded)

Development of Recommendations

The identification of issues and recommendations of "starting points" are the subject of the next two chapters: a summary of the discussions and a set of recommendations. Synthesis of these ideas occurred after the conference ended. Because of the extensive notes taken during the conference (thanks to the availability of about half a dozen laptop computers and to the expertise of conference participants at being able to engage in discussion along with taking notes), this process was nourished by an abundance of excellent ideas. Reading through the notes was almost like entering into each discussion group again.

Publications and Dissemination

This volume of conference proceedings is the primary record of the conference. In addition to this description of the design of the conference, it contains a summary of the discussions, recommendations for future action, all of the working papers, and supplementary information.

An executive summary is also available for wide dissemination. For example, we hope that the National Science Foundation will mail this summary to all authors of preliminary proposals for NSF's Teacher Enhancement cycle of September 1, 1998. We have also made it available to personnel in the state departments of education, National Council of Supervisors of Mathematics (NCSM), Association of State Supervisors of Mathematics (ASSM), and Association of Mathematics Teacher Educators (AMTE). Announcement of the availability of this proceedings book and the executive summary have also been posted in

(a) the listserv of American Educational Research Association (AERA), (b) newsletters and listservs of National Council of Teachers of Mathematics (NCTM), Research Council for Diagnostic and Prescriptive Mathematics (RCDPM), National Science Teachers Association (NSTA), relevant special interest groups and divisions within AERA (e.g., mathematics education, science education, assessment, teacher development, etc.), Association of Teacher Educators (ATE), Association of Colleges of Teacher Education (ACTE), (c) ERIC Mathematics and Science Education Clearinghouse, and web sites such as North Carolina Department of Public Instruction and UNC Mathematics and Science Education Network. Other publications (e.g., a "teacher-friendly" discussion of classroom assessment) are in progress, but they are not complete at the time of the writing of this book.

There are several audiences which need to be targeted for dissemination of the products of this project. "Obvious" constituencies include state mathematics supervisors (e.g., ASSM), district mathematics supervisors (e.g., NCSM), mathematics researchers and teacher enhancement project directors (e.g., SIG/RME), directors of SSI/USI/RSI projects, and mathematics teacher educators (e.g., AMTE). In addition, it seems advantageous to be sure that some "non-mathematics educators" also have access to these products so that recommendations can be implemented across disciplines. Non-mathematics educators include key state or district superintendents, state or district assessment directors, state or district science education supervisors, and nationally known staff development specialists.

There are also plans being made (at the time of the writing of this book) for a dissemination conference to be held in conjunction with the 1998 NCTM annual meeting. The one-day conference is planned for Wednesday, April 1, 1998, immediately prior to the start of the NCTM annual meeting on Thursday, April 2, 1998. Wednesday is positioned immediately after the end of the ASSM annual meeting, and it is coincident with part of the NCSM annual meeting and the research presession co-sponsored by NCTM and the SIG/RME. The dissemination conference will include an overview of the discussions from the working conference and discussion of the recommendations. Audience members will have opportunities to express their opinions concerning both the discussions and the recommendations.

References

Briars, D. (1996 July 22). *Assessment Communities of Teachers project* [interview]. National Science Foundation Teacher Enhancement Project.

Carpenter, T. P., Fennema, E., Peterson, P. L., Chiang, C-P., & Loef, M. (1989). Using knowledge of children's mathematics thinking in classroom teaching: An experimental study. *American Educational Research Journal, 26*, 499-531.

Darling-Hammond, L., Ancess, J, & Falk, B. (1995). *Authentic assessment in action: Studies of schools and students at work.* New York, NY: Teachers College Press.

Driscoll, M. (1995, May). "The farther out you go...": Assessment in the classroom. *Mathematics Teacher, 88*, 420-425.

Fennema, E., Carpenter, T. P., Franke, M. L., Levi, L., Jacobs, V. R., & Empson, S. B. (1996). A longitudinal study of learning to use children's thinking in mathematics instruction. *Journal for Research in Mathematics Education, 27*, 403-434.

Fennema, E., Carpenter, T. P., & Peterson, P. L. (1989). Teachers' decision making and cognitively guided instruction: A new paradigm for curriculum development. In N. F. Ellerton & M. A. Clements (Eds.), *School mathematics: The challenge to change* (pp. 174-187). Geelong, Victoria, Australia: Deakin University Press.

Fennema, E., Franke, M. L., Carpenter, T. P., & Carey, D. A. (1993). Using children's mathematical knowledge in instruction. *American Educational Research Journal, 30*, 555-584.

Hembree, R., & Dessart, D. J. (1986). Effects of hand-held calculators in precollege mathematics education: A meta-analysis. *Journal for Research in Mathematics Education, 17*, 83-99.

Hembree, R., & Dessart, D. J. (1992). Research on calculators in mathematics education. In J. T. Fey & C. R. Hirsch (Eds.), *Calculators in mathematics education: 1992 yearbook* (pp. 23-32). Reston, VA: National Council of Teachers of Mathematics.

Hiebert, J., & Carpenter, T. P. (1992). Learning and teaching with understanding. In D. A. Grouws (Ed.), *Handbook of research on mathematics teaching and learning* (pp. 65-97). New York: Macmillan.

Kulm, G. (1990). New directions for mathematics assessment. In G. Kulm (Ed.), *Assessing higher order thinking in mathematics* (pp. 71-78). Washington, DC: American Association for the Advancement of Science.

McConnell, J. W., Brown, S., Eddins, S., Hackworth, M., & Usiskin, Z. (1990). *The University of Chicago School Mathematics Project: Algebra.* Glenview, IL: Scott, Foresman and Company.

National Council of Teachers of Mathematics. (1991). *Professional standards for teaching mathemtaics.* Reston, VA: Author.

Nicholls, J. G., Cobb, P., Yackel, E., Wood, T., & Wheatley, G. (1990). Students' theories about mathematics and their mathematical knowledge: Multiple dimensions of assessment. In G. Kulm (Ed.), *Assessing higher order thinking in mathematics* (pp. 137-154). Washington, DC: American Association for the Advancement of Science.

Peterson, P. L., Fennema, E., Carpenter, T. P., & Loef, M. (1989). Teachers' pedagogical content beliefs in mathematics. *Cognition and Instruction, 6,* 1-40.

Rachlin, S. L., Matsumoto, A. N., & Wada, L. A. T. (1992). *Algebra I: A process approach.* Honolulu, HI: Curriculum Research and Development Group, University of Hawaii.

Resnick, L. B. (1987). *Education and learning to think.* Washington, DC: National Academy Press.

Romberg, T. A., Zarinnia, E. A., & Collis, K. F. (1990). A new world view of assessment in mathematics. In G. Kulm (Ed.), *Assessing higher order thinking in mathematics* (pp. 21-38). Washington, DC: American Association for the Advancement of Science.

Silver, E. A., & Smith, M. S. (1996). Building discourse communities in mathematics classrooms: A worthwhile but challenging journey. In P. C. Elliott & M. J. Kenney (Eds.), *Communicating in mathematics, K-12 and beyond: 1996 yearbook* (pp. 20-28). Reston, VA: National Council of Teachers of Mathematics.

Smith, J. P., III. (1996). Efficacy and teaching mathematics by telling: A challenge for reform. *Journal for Research in Mathematics Education, 27,* 387-402.

Steffe, L. P., & D'Ambrosio, B. S. (1996). Using teaching experiments to enhance understanding of students' mathematics. In D. F. Treagust, R. Duit, & B. J. Fraser (Eds.), *Improving teaching and learning in science and mathematics* (pp. 65-76). New York, NY: Teachers College Press.

Webb, N. (1996). *Evaluation study of Interactive Mathematics Project.* Madison, WI: University of Wisconsin, Wisconsin Center for Education Research.

Worthen, B. R. (1993, February). Critical issues that will determine the future of alternative assessment. *Phi Delta Kappan, 74*(6), 444-454.

Appendix A: Student's Comments

Ladies and gentlemen, my beloved teachers, and honored guests, I am honored to be selected to speak to you this evening about my experience as a high school student in Wake County. My name is Philip Salib. I am graduating this year from the Leesville Road High School and will pursue my education at UNC - Chapel Hill. I am an honor student with a GPA of 4.33. My extra curricular activities take up much of my time. I am in the school band, for which I serve as the business manager, tracking activities with a budget of more than $90,000. Additionally, I attained the rank of Eagle Scout with the Boy Scouts of America, and I am active in my church's youth group. As you can tell, I have done well academically and socially, though my parents tell me that I can and *should* do better.

Recently, my school had Congressman David Price as a guest speaker at the honors recognition ceremony. In his speech, Congressman Price told us that we did not accomplish our successes by ourselves and that we got to where we are with the help of our parents, friends, and *teachers*. I completely agree with that statement. I was fortunate to have been a student in an excellent school, in an excellent school system, taught by excellent teachers, and have excellent parents.

Now I would like to give you some feedback about certain assessment strategies that I considered helpful and others that caused some difficulties.

Having teacher response regarding my overall performance is very important in my education. When the teacher tells me how I am doing in the class, it helps me to better prepare myself. If the teachers tells that I am doing fine, then I work hard to maintain my good status. If the teacher tells me that there are certain things that need to be given more attention to in my studies, then I try to spend more time to improve these weaker areas. The best way for a teacher to convey his/her ideas to the student is to take some time to write some feedback on some of the tests or quizzes that are returned to the student.

Feedback from the teacher is not only important in academics, but also in class attitude. I would find it very helpful if the teacher would inform me on what types of attitudes he/she likes from me and also what type of attitude he/she dislikes. To me, it seems this type of feedback would better improve the environment of the class and would make the class a better place to learn. The best way for the teacher to convey these ideas is to have the teacher verbally tell the student what he/she is thinking either before or after the class.

All of my teachers have been excellent and very essential to my education (especially in math). Throughout my years in high school, my math teachers have probably been the most helpful in terms of teaching and being able to explain the information thoroughly. Mrs. Kolb is my current math teacher for AP Calculus. She has done a *spectacular* job. It is more important to her that her students are enjoying math rather than plowing through the material. This makes learning more fun and makes an extremely difficult subject seem not so hard. She is very familiar with the subject she is teaching. I commend her on a job well done and thank her greatly for the wonderful learning experience she has allowed me to have.

On the other hand, I have had teachers who taught straight from their notes. These teachers were unable to answer questions outside of their daily plans. The worst thing for a student is to have an unanswered question and not have the material explained completely. On this note, I would like to offer some suggestions to better the learning environment:

1. *Resources*: A good book and a good set of teacher's notes are very helpful. The most difficult thing for me was to go home, try to do my homework and find out that the book doesn't cover the material and the teacher just brushed on the subject and I have to waste time on researching information.

2. *Homework*: Homework is an effective way of reinforcing what we learn in class. That's why the homework should cover every possible concept related to the subject being taught. The more, the better. Busy work is not beneficial in any way. It simply wastes valuable time. It is also, important that the teacher would review with the class the homework, so everyone would know if they did it correctly and in case someone didn't know the answer to one or more problems. The best way, in my educational experience, to review the homework is to have the students put them up on the board. Sometimes it is easier for other students to explain certain topics rather than having the teacher continually trying to reinforce the material.

3. *Testing*: Tests should first measure how much of the concepts were absorbed. Challenging questions should provide extra credit and little or no penalty for not being able to answer them.

4. *Grading*: The grading system should be forgiving to a person who messes up on one of the tests. This can be achieved by ignoring the lowest test grade. There are many reasons why a student would perform poorly in a test in spite of the fact that he knows the subject. That could be because of not feeling well at that time because of health, psychological or social reasons.

One very important factor in grading papers is the speed in which they are returned. The faster the tests or quizzes are returned, the more effect the grade has on the student. The material is also fresh in the mind of student. Some teachers procrastinate in returning papers. When the student gets these grades back, they don't seem to care as much about their grade and most importantly, their mistakes.

Also, there is some experimentation with group averaging grading system. This should not be permitted, because while it helps poor performers, it is unfair to the excellent performers. Proponents of this system may claim that it teaches team work, but let us not forget that individuals should always be responsible for their own performance.

5. *Activities that waste time*: Those are activities such as science projects and other projects that require arts and crafts. Those projects consume a lot of time in obtaining materials and performing labor which does not contribute to teaching of the subject. Science projects should be optional for rewards and credit other than grades. Science experimentation should be limited to the lab, if actual knowledge and lab skills are the goals. Projects should be carefully selected and carefully explained to the student. It is very bad that a student should work hard on a project and have absolutely no clue as to the relevance of the project to the subject they are studying.

I must say that all of my teachers have been commendable and should all be given raises. At my school, with the teachers I have had, there is not much you can improve on. I must, again give great thanks to all my teachers, especially Mrs. Kolb, for the wonderful times I have spent in their classrooms.

I greatly thank all of you for giving me the chance to share some of my thoughts and hope my comments have been helpful to you all. Thanks again and I hope to see some of you in the future.

Appendix B: NSF Representative's Comments

We are living in interesting times for mathematics assessment! Quite regularly, mathematics test scores make the headlines in *USA Today*, or are the subject of a commentary on NBC News. Reports of NAEP scores in mathematics, results of the TIMSS tests, and the President's announcement of a new National Mathematics Test for the eighth grade have been subjects of much discussion at professional meetings in the community. At the recent Joint Mathematics Meeting in San Diego last January, attendance at a session led by Jim Stigler on the TIMSS results and video filled a large hall. And "ditto" for similar sessions at NCTM, the TIMSS Symposium here in Washington, etc. There's a lot of attention on student achievement data in mathematics these days.

High stakes assessment has its place. The results can serve as benchmarks to help us gauge how we're doing. But, high stakes assessment is but one part of the assessment spectrum. A larger piece of that spectrum is what goes on, day in and day out, in mathematics classrooms at all levels. It is classroom assessment that is the focus of this conference.

There are just three points I want to make about classroom assessment:

1. Classroom assessment takes many different forms, all the way from those that are informal and embedded to those that are quite formal and summative in nature. Every time a teacher asks a student a question and the student responds, assessment is (or should be) going on. That teacher has to make a quick assessment of the validity of the student's thinking and, in a Standards-based classroom, frame another question that pushes that student and others in a direction that is mathematically productive. Good teachers have always engaged in a lot of informal classroom assessment. Assessment has been around as long as teaching and learning, but what's different is that we're now formalizing -- and perhaps making more rigorous -- our questioning strategies and our accompanying assessment strategies.

2. Newer styles of assessment place great demands on the classroom teacher in the areas of content knowledge. Teaching in a Standards-based classroom means that students take a more active role in their learning, and this creates opportunities for students to question/respond in ways that challenge the traditional unfolding of

classroom mathematics. The more a teacher moves in directions that promote student investigation in the classroom, the more knowledgeable the teacher must be about mathematics.

3. Professional development that centers around student assessment in mathematics offers the potential for teachers to learn a great deal about the discipline, its aesthetics, and its values. How do you decide in a scoring rubric that a certain type of student response merits a score of "3" and another type of response a "4" without reflecting on the validity of the justification? On the language of mathematics? Even on its aesthetics (has the student, for example, come up with a solution or argument that's "slick"? and what does it mean to be "slick"?). In other contexts, "slick" is not very desirable (how many of you want to buy a car from a "slick" salesman?), but, in mathematics, a "slick" proof is "cool"!

We are very pleased to provide support for this conference. We look forward to the discussions to come and to the recommendations you will make on this important topic of classroom assessment in mathematics.

Understanding and Improving Classroom Assessment: Summary of Issues Raised

George W. Bright
The University of North Carolina at Greensboro

Jeane M. Joyner
North Carolina Department of Public Instruction

The discussions during working group sessions were vital to the work of the conference. To encourage interaction, working group composition was changed each day. In this way, each participant was able to interact with almost every other participant at some point during the conference.

During each working group session, participants volunteered for one of several roles: facilitator, recorder, and reporter. The recorders used laptop computers to make on-going notes during the working group discussions. During the reporting sessions that followed the working group sessions, some individual participants also volunteered to amplify on particular positions or comments they had made during the discussions. The summary in this chapter is intended to capture the major points that were either outlined during the reporting sessions, embedded in the group discussion notes, or presented in the final versions of the working papers (completed after the conference was over), while at the same time trying to retain the diversity of the voices that were heard during the conference. There was no effort to come to consensus in the time allotted for discussions, so many comments reflect either individual or small group positions.

One important caveat needs to be stated at the outset. Classroom assessment is not the same as grading. Although assessment in general is often thought about in the context of "grading," this conference really focused on the type of assessment whose main purpose was to inform better instruction and to communicate to students and others what students "really know" about mathematics. This focus was explicitly imposed in part by the charge to participants. As is evident in what follows, participants were collectively quite aware of the complexities of classroom assessment and of the various purposes for different kinds of assessments.

The summary is organized around seven main sections:

- Defining classroom assessment and clarifying assumptions
- Interaction with external assessments
- Connecting classroom assessment to instruction
- Teachers' knowledge base in implementing better classroom assessment
- Supporting teachers in implementing better classroom assessment
- Special concerns: technology, equity, and preservice education
- Research needed in support of effective classroom assessment

Throughout this chapter, reference is made, denoted [Author], to the working papers. These papers provide additional detail about many of the points made. We have tried to include critical main points from these papers, without being repetitious of the details.

Defining Classroom Assessment

Assessment, in general, has been described as a process of gathering evidence and making inferences about students' learning [Webb]. In the words of a famous song writer, "I wish I could get behind your eyes and see the world as you see it." That is, one goal of classroom assessment is to understand learning from each student's point of view.

It was pointed out in the discussions that all teachers already "do" classroom assessment in some form. Most of the conference discussion, however, focused on what a "reform image" of classroom assessment might look like and how teachers could be supported as they moved toward that admittedly vague vision [Fennell; Rachlin].

Classroom assessment involves monitoring and making decisions. Classroom assessment differs from other assessment in that it is more focused on using the inferences to inform instruction and to monitor day-to-day progress [Joyner]. Classroom assessment might be described as the gathering and interpreting of evidence about students' learning in order to make decisions about instruction within a classroom. It seems important to keep classroom assessment focused on individual classrooms. Classroom assessment is what individual teachers do on a day by day basis so that they can improve instruction. It is an on-going process of collecting information on a class and individuals in that class so that a teacher understands where those students are in the context of mathematical development.

Teachers must first identify what they are looking for. This means that teachers cannot assess without being grounded in both understanding of the content they are teaching and what parts of that content they value. Instruction can be informed by classroom assessment both in the very short term (How should I change the instruction right now? What problem should I pose tomorrow?) or in a much longer term (How should I plan the next unit to help students build on what they already know?).

Classroom assessment might be thought of as taking the temperature of learning or judging the climate of instruction. It is important to know what students think and understand. It is important for teachers to know when students have something under their belt (akin to "mastery") and are ready to go on to the next "thing." Good teaching involves knowing when a mathematical idea is integrated in a students' thinking.

One of the main characteristics of current conversations about classroom assessment is that it shifts attention away from telling (i.e., what the teachers says) and more toward understanding (i.e., what the students learn about mathematics). In order to make this shift, teachers need to gather a wide range of information about the mathematical understandings that students are developing and then learn to interpret that information [Briars]. Because of the importance of understanding students' mathematical thinking, the conference discussions focused mainly on assessment of students' content knowledge.

Conference discussions often seemed based on the implicit assumption that one of a teacher's main jobs is to help students build understanding. Classroom assessment, because of its focus on information about what students understand, helps prevent "illusions of learning." Teachers often seem to want so much for their students to understand, that they tend to over-generalize the depth of understanding that is actually present in what students say and do. Teachers often infer more about what students understand than the evidence provided by students' performance and explanations really supports. Sometimes this happens because teachers imbue the words that students use with all of the same understanding that the teachers themselves would mean with use of those words. Because of this over-generalization, teachers are sometimes surprised when students do not solve problems or do not score well on external assessment. Teachers often have created in their own minds an illusion about students' having deeper knowledge of mathematics than is actually the case.

Classroom assessment can be conceptualized as an on-going process that uses learning frameworks, standards, and goals to formulate questions and tasks that reveal what students know and can do [Jenkins]. Classroom assessment helps teachers know if students really understand mathematics, rather than just giving the appearance of understanding. There are four primary elements: preparation, classroom interaction, reflections, and consequences.

- *Preparation* includes the identification of goals of mathematics instruction, both as part of general curriculum development and as part of lesson planning by individual teachers in particular classrooms. Indeed, what is assessed is highly influenced by what is viewed to be important, as specified in these goals. But beyond this, preparing requires that teachers understand mathematics and internalize background information about (a) an understanding of the thinking (ideas and strategies) that underlies students' development of mathematics topics, (b) an awareness of strategies, including questions and tasks, that reveal this thinking (beyond, for example, memorized procedures), and (c) the scope and sequence of the mathematics content, including understanding of what comes before and what comes after the content being taught. Teachers then have to adapt this knowledge to fit the particular group of students in a class.

- *Classroom interaction* refers to the ways that teachers and students engage together in tasks and discourse. Students and teachers need to monitor progress informally on a daily basis as individuals and in large and small groups. In order for interactions to be effective, teachers, students, curriculum authors, and others must be committed to the goal of exposing students' mathematical thinking to public scrutiny. Interactions help teachers focus on what students think (e.g., reasoning), what they know (e.g., concepts), and what they do (e.g., procedures). Critical teacher behaviors in this process include listening, observing, and interpreting. It is during this phase that information is gathered by teachers in multiple formats.

- *Reflections* provide an opportunity for teachers to make sense of what students are doing and for students to make sense of their personal understanding. It is primarily during reflection that teachers place students within frameworks organized around current understanding of students' mathematical development and determine how well students are progressing toward the instructional goals. These are inferences made from information gathered in the earlier phase of the process [Briars]. Reflections help teachers decide if instruction is being effective so that changes and modifications can be made. During reflections, teachers decide what to do next, which students to push, and so on. Some of the reflections will be formative "evaluation" of students' progress, and some will be summative "evaluation" that compares students' progress against established standards of performance.

- All classroom assessment has *consequences*, the most important of which is more effective, quality instruction. Teachers make instructional decisions based on their knowledge of what students know and can do. In addition, teachers may assign grades, conference with students individually to share what the teacher has learned, or report students' progress to adults outside the classroom. Whatever is reported, however, needs to be linked to the goals of instruction that were identified during the preparation phase. The daily instructional decision making may not be communicated regularly to outsiders; these decisions are "simply" part of the job of teaching.

This view of assessment points out a distinction between assessment and instruction. Reflection and consequences are essential parts of classroom assessment; they may not be seen as clearly as essential parts of instruction. However, highlighting the links between assessment and instruction may be an important part of clarifying the nature of classroom assessment.

Perhaps the central notion of classroom assessment is that teachers use many techniques and strategies to build a model of a student's thinking [Berenson, Vidakovic, & Carter; Moss; L. Williamson]. That model is built within the context of a development scheme, and then teachers make conjectures about what tasks and experiences students should have to move them along that developmental path. Teachers want their students to learn. However, without internalization of appropriate developmental frameworks, teachers might not be able to interpret what a student knows and what the next appropriate task might be.

It appears that many teachers "assess to assess" (i.e., assess to assign grades); relatively few teachers assess to learn about their students' thinking. For many reasons, assessing for instructional decision making is not part of most teachers' reasons for assessing. For example, many teachers do not recognize the role of classroom assessment in making instructional decisions, preservice education may not have prepared teachers to assess for instructional decision making, and curriculum materials do not typically provide adequate assistance for teachers in using classroom assessment in instructional planning. One of the primary goals of classroom assessment, however, at least within the mathematics education reform movement, seems to be to learn about students' understanding as a means of informing instructional decisions.

Classroom assessment is a different "take," however, on what teachers know and can do. Classroom assessment helps teachers reorganize their knowledge of students, the curriculum, pedagogical practices, and possibly even mathematics itself. Such reorganizations are not likely to occur unless there are communities of teachers (e.g., discussion groups set up in the MathLine project) which collectively develop techniques for understanding students better. Those communities cannot do their work unless there is time for them to reflect together about what they are learning about students. Schools may need to be restructured to make time for such reflection as well as to assure that teachers get consistent messages (e.g., from curriculum, external assessments, teacher evaluation practices) that understanding students is important.

There are many filters that influence what information teachers might decide is important to gather. One critical filter is the teacher's content knowledge. The need for teachers to have fairly deep understanding of mathematics was a theme often repeated in the conference discussions. Other filters include teachers' views about the nature of mathematics, teachers' beliefs about the capabilities of their students, the nature of the curriculum being used, and teachers' perceptions of community expectations.

Assessment should provide students with multiple avenues of access. Teachers need multiple sources of information in order to make reliable inferences about students' learning, but students also need multiple avenues for demonstrating the full range of what they know. Having all students strive for high standards means that there are likely to be several paths to the same goal and several ways to demonstrate achievement.

The inferences that teachers can make are limited by, among other things, what students are willing or able to reveal about their thinking. It is important, therefore, to have a classroom climate where students feel safe revealing both strengths and limitations of what they know. Teachers who can make good inferences about their students' thinking (i.e., make sense of what they hear students say and what they see students do) seem more likely to help students grow in mathematical understanding [Lindquist; Rachlin; Webb].

In general, learning is sense-making. Learning to teach is, at least in this respect, no different. Teachers need to make sense of mathematics, the curriculum, and their students' thinking. Assessment, in general, and classroom assessment, in particular, can inform this process of sense-making about students [Jenkins]. Teachers' "intentionality" with respect to assessment is important for sense-making. Teachers should plan how they will listen to students, develop tasks, make decisions, etc. Quality classroom assessment seems unlikely to be created when teachers simply let these things "just happen." This is not to say that classroom assessment does not happen "on the fly," but rather it points to the need for teachers to plan consciously for what and how they will assess.

Although some teachers seem to be good at making sense of students' thinking, it seems difficult to pass this capacity on to other teachers. Just as students need to develop understanding of mathematics, teachers need to develop understanding of classroom assessment. Constructing understanding is always a struggle, but the struggle to understand classroom assessment seems amplified because of the complexity involved.

Classroom assessment provides perhaps the only known way for teachers to make curriculum fit the particular students in a class. Every set of curriculum materials is written for the "generic student;" no curriculum can fit the students in every classroom. Effective adaptations of curriculum need to be built on knowledge of what students in a particular class understand and can do. Classroom assessment seems to be the most effective way for this information to be generated.

Classroom assessment appears to represent a fairly low cost vehicle (at least in terms of materials and equipment) for reform of mathematics instruction. It may not be so low cost, however, when the costs of inservice for teachers and reflection by teachers are taken into account. As teachers begin to reflect more deeply on the understanding of their students and the effectiveness of their instruction for developing better thinking by students, then teachers may "naturally" take on more responsibility for change in their own classrooms. Thus, the return on this investment is potentially quite profitable.

Mathematics and classroom assessment. Essential mathematical ideas must be at the core of any assessment task, and anyone (including, but not restricted to, teachers) who designs assessment tasks must understand the "heart" of mathematics as a discipline [Parker; Spresser; Thompson; Zawojewski & Silver]. The assessment tasks that a teacher selects will send clear messages to students in the class, and others outside the class, about what parts of mathematics are important to learn. For this reason, it is important to engage teachers and others in conversations that will help them to understand explicitly what messages different kinds of tasks send about the domain of mathematics [Rachlin].

One widely held view among conference participants was that assessment tasks must be consistent with the idea that mathematical thinking is a sense-making activity [Clements]. One working group commented that assessment targets must be both important and robust mathematically; that is, mathematical ideas that help organize conceptual understanding and that make application of mathematical understanding easier [Spresser]. If assessment tasks do not mirror these characteristics, we may find ourselves holding students and teachers accountable for the wrong things. At the same time we have to educate policy makers (including legislators) that sense-making is probably the single most important characteristic of quality assessment.

One of the main reasons for highlighting sense-making is that learning goals should not be limiting for students. Good assessment will support all students in accomplishing the task in different ways, and the strategies that students choose to use in accomplishing the task reveal much about the level of understanding of students. Being able to complete assessment tasks in different ways also sends the message to students that personal understanding of the mathematics content is critical.

Conference participants certainly recognized that content knowledge was not the only aspect of mathematical knowledge that was important. Students' disposition and attitude are also important, particularly as they are manifest in confidence and perseverance. Without confidence and perseverance, students are not likely to persist long enough to solve meaningful problems [Bush]. There was little time during the conference, however, to list ways that teachers might use to assess these affective characteristics of students.

During the conference, several examples were offered of ways to change assessment tasks; some of these came from, or were based on the principles in, the *Assessment Standards* (NCTM, 1995) and other documents. Instead of asking students to compute 347 - 196, ask, "What two numbers have a difference of 151?" This will allow students to indicate the kinds of numbers they are comfortable with. Instead of giving students a plane figure with all sides labeled and asking them to find the perimeter, give them the value of the perimeter and ask them to create one or more figures with five sides with that perimeter. This will allow them to indicate how they break apart an idea into its component parts.

Planning classroom assessment. When teachers plan any classroom assessment, they must consider many different questions. How will they be certain that students clearly understand the learning targets? What are the criteria [Briars; Sowder; Thompson] for good work? What strategies might students use in completing assessments? How might students communicate what they are learning? What are strategies to involve students in goal setting and self assessment? What assessment strategies will give the most information about students' achievement of the learning targets? How can mathematics be embedded in a variety of contexts during assessments?

Teachers do have a variety of classroom assessment strategies to use: portfolios, interviews, questioning, journaling, projects, etc. These strategies are also becoming increasingly familiar to teachers. But it seems likely that many teachers do not know enough about the differences in the quality of information that these different strategies generate [Sowder].

One of the key questions for teachers is deciding what to assess [Richardson; Spresser]. Teachers have to decide what they think is the most important mathematics for students to learn. One approach is identifying the "big ideas" in mathematics. For example, fluency with numbers seems to be one of the big ideas across much of mathematics instruction. But teachers also have to decide what it means for students to "understand" mathematics.

It is also important to consider whether the types of effective classroom assessment strategies change as teachers gain experience. Certainly the range of strategies will increase during a teacher's career, but there may be some "elementary" strategies (e.g., asking the questions of "Why?" and "How? ") that yield information that is useful to beginning teachers and "advanced" strategies (e.g., using different questions to probe a student's ability to communicate versus probing a student's ability to understand a mathematical process) that yield information that is useful to more experienced teachers. Even experienced teachers who are changing their classroom assessment strategies may find that early in the change process one type of strategy is effective, whereas later in the change process a different type of strategy is effective. Much more needs to be known about the process of development as an effective classroom assessor.

Communicating expectations and outcomes of assessment. Part of every implementation of classroom assessment is the communication of expectations (i.e., feedforward) and progress toward goals (i.e., feedback). Communication with students [Sowder] can proceed in any of the following ways:

- Show students standard-setting performances and have them discuss why they are high quality.
- Show and discuss the criteria used to judge performance.
- Have students judge each other's work using criteria (e.g., in the form of a rubric) and provide feedback to each other.
- Allow students to be part of the judging/standard-setting process as early as possible in their education within and across grades.
- Give students tools and opportunities to become self-assessors.
- Allow students to gain ownership of expectations through goal setting.

- Allow students to monitor their own progress by posting their problem solving, writings, etc., so that performance is shared among all students.
- Talk to the students and have students talk to each other about their understanding. Indeed, one job of teachers after *they* listen to students is to help students listen to each other.
- Give feedback, frequently and in a timely manner, on what students can do to improve their performance.

Similarly, the importance of classroom assessment can be communicated to teachers, administrators, policy makers, and families, in many ways [Lindquist; Midgett; Parker; Sowder]. In each of these scenarios, it will be important to make the case forcefully that teachers need resources (e.g., time, money, and training) in order to move toward better classroom assessment. Educators and non-educators alike need to understand the influence of policy makers on the implementation of quality classroom assessment.

Informing various policy makers and stakeholders about the power and untapped potential of classroom assessment can take several forms:

- Have stakeholders experience assessment tasks and then discuss the mathematics knowledge and thinking demanded by those tasks (putting stakeholders in the role of students).
- Have stakeholders look at student work and rubrics for judging that work. Student work shows not only the value of the assessments but also the level of accomplishment of the students.
- Show results from past performances on other assessments and explain how classroom assessments make a difference in the level of students' performance. For example, a video that shows the interactions of the teacher with students might constitute a convincing argument for the value of classroom assessment.
- Have groups make judgments about students' thinking based on different kinds of evidence generated from different kinds of assessment tasks. (That is, put people in the role of assessors.) Then provide one or more contexts within which to make these judgments.

- Discuss case studies, ethnographic studies, and so on, so that everyone understands the role of researchers (including the teacher as researcher).

Both students in the classroom and adults outside the classroom need to understand that implementing effective classroom assessment takes time. Teachers need time to learn how to carry out classroom assessment. There may be a need for teachers to reorganize their time within the instructional day to implement the techniques they have learned. Classroom assessment should be integrated into instruction so that is not simply "one more thing to do" during instruction. Implementation of effective classroom assessment imposes a responsibility on the teacher to be sure that instructional time is spent profitably. It also requires a high level of trust among students and adults outside the classroom that teachers will choose assessment tasks that are aligned with the curriculum and that are equitable to all.

Beyond communicating with students about the expectations and goals of instruction, it is probably important to involve students in the assessment process itself. Students need to be as much a part of the assessing as the teacher. Students should learn techniques for understanding their own learning. They need to learn to value the development of conceptual understanding of mathematics rather than just procedural knowledge.

Teachers need to create a climate that fosters students' taking an active role in the process of evaluating understanding [Parker; J. Williamson]. Within such a climate, students will come to believe that it is important that mathematics makes sense and can be useful in dealing with the world. Students should learn to ask themselves generic questions (e.g., "How did I arrive at my answer?") but teachers will also need to help students develop other questions that might be related to specific mathematics ideas.

Teachers will need to help student clarify learning targets and make judgments about how far away they are from those goals. Ultimately, students might be asked to choose a particular method of assessment that they think would reveal the most about their thinking, but this level of student involvement would probably require rather sophisticated knowledge of mathematics and of classroom assessment by both the teacher and students. It seems likely that intimate involvement of students in the assessment process will happen only if teachers know and understand a great deal about the curriculum, about students' thinking, and about various ways to assess.

Conference participants shared many anecdotes of students' learning to "self assess." Some students do develop a good sense of what they know and what they don't know. One example that seemed to resonate well was of one primary-grade student who wrote, "If I were 100 years old, I would go to a nursing home. I would stay there until I was dead. By the time I was 100, I would know regrouping with subtraction and then I would die happy."

Interaction with External Assessments

All parties (teachers, students, parents, school administrators, legislators, etc.) must understand the different uses that are intended to be made of different kinds of external assessments. They must demand that any external assessments which are imposed for evaluation of the implementation of state curriculum guidelines (e.g., end-of-year tests) should be carefully aligned with those curriculum goals as well as with instructional practices in the classroom [Kenney; Lindquist; Muri; Parker]. Other external assessments, because they are "national" in scope or because they are used to rank students nationally, may intentionally ignore alignment with particular curriculum guidelines.

Classroom assessment practices, because they are tied very closely to individual classrooms, are likely to be quite variable across classrooms. Many external assessments, in contrast, are designed to provide uniformity across many classrooms. External assessments, then, may not reflect either the learning goals or the classroom practices of any particular classroom.

When an external assessment is well-constructed (e.g., worth teaching to), it asks students to engage in conceptual thinking about mathematics. It also helps them "unpack" their thinking (e.g., understand the strengths and limitations of what they know). The act of unpacking may itself help refine conceptual thinking [Kenney]. In particular, external assessments need to include evaluation of understanding of concepts and processes as much as (or, perhaps, more than) mere facility with facts and skills. While it is wrong to "teach the test," it is irresponsible not to "teach to the test," especially if the test measures what we value.

It is important to think carefully about why external assessments are mandated and who benefits from the knowledge they generate about students' performance. Legislators, school administrators, and parents all gain some information from external assessments, but it appears that the main purpose may be in some sense to "justify" school expenditures, particularly through monitoring program implementation. Another frequent use of some external assessments (e.g., SAT), however, is to rank order students. Any assessment that does this is built on particular values, especially with respect to identifying what constitutes mathematical reasoning. People who use these rankings of students (e.g., university admissions officers) need to understand the value system inherent in the assessment.

Many people (e.g., teachers, administrators, students, parents) are frequently concerned about the alignment of external assessments with the school curriculum, and hence, with classroom assessment practices. It is generally acknowledged that external testing (local, state, or national) can influence instruction and consequently things that happen in the name of classroom assessment [Muri]. Teachers appropriately feel an obligation to help their students learn to succeed on external assessments, so teachers are likely to feel pressured to make their classroom assessments "look like" or at least look similar to the external assessments. In order to do this, teachers need to become familiar with the types of questions included on these external assessments [Kenney]. However, because the purposes and audiences for results of external assessments are not the same as those for classroom assessments, these responses by teachers may actually interfere with teachers' understanding of their students' mathematical thinking. That is, external assessments can distract teachers from the goal of understanding their students' thinking, at least in part by limiting the range of assessment strategies that teachers may consider appropriate.

One other concern about external assessments is the extent to which schools can hold onto (or be required by legislatures to hold onto) existing assessments while simultaneously promoting reform. Many participants expressed skepticism about the "chilling effect" on reform of using only traditional external assessments [Kenney]. There appears to be a need to make a shift in the nature of external assessments along with making shifts in instruction.

Teachers may be able to connect external assessments to learning targets and to understanding students' thinking through discussion with students of the results of the external assessments (e.g., standardized tests, portfolios). Such discussions might help students learn how to connect those assessment tasks to what they know. In order to do this effectively, teachers would likely need to become quite familiar with the nature of external assessments [Zawojewski & Silver].

There were several concerns raised about the equity of some external assessments. Clearly, the level of English proficiency of students may influence their performance. Too, collecting and interpreting data (e.g., by a state department of education) without understanding the characteristics of the students who completed the assessment may generate inequities through inaccurate interpretations of performance.

One view that surfaced during the conference was that if classroom assessment were operating at an optimal level in every classroom, there might not be a need for externally-mandated testing at any level. Underlying this is the belief that if students are able to solve problems and explain their thinking (as would be required in any implementation of a reform vision of classroom assessment), they are ready for "any" external assessment. Scores on external assessments "will take care of themselves." It seems somewhat doubtful, however, that teachers in general, or the general public, would agree with this view.

One clear message coming from the conference discussions was that external assessments should not drive what happens related to on-going classroom assessment. Teachers use the information from classroom assessment for different purposes than the purposes of most external summative assessments. Teachers should remain committed to using classroom assessment for the more immediate purpose of planning instruction.

Connecting Classroom Assessment to Instruction

Most participants would seem to agree with the proposition that instruction and classroom assessment should be tightly connected. If instruction were rich and varied, then it would almost "naturally" be accompanied by classroom assessment strategies that would also be rich and varied. Teachers would have an extensive repertoire of ways to look at student progress [Fennell; Gunter; Rachlin; Thompson]. It is not clear, however, how (or when or whether) changes in classroom assessment might require changes in instruction.

There was some divergence of opinion, however, about whether teachers should plan instruction first and then plan assessment or whether planning for instruction and assessment ought to proceed together. There did appear to be consensus that teachers should identify the mathematics to be learned prior to designing any assessment (or instructional) task.

Information generated through classroom assessment should help teachers understand the impact of their teaching and should suggest changes that teachers might want to make in instruction [Briars; Bright; Moss; Muri; Spresser; L. Williamson]. What students know and do not know seems to be appropriate evaluative information about the quality of teaching. After all, one of the fundamental reasons for classroom assessment is to answer the basic question, "Based on what you learned about your students' thinking today, what will you do tomorrow?"

It seems reasonable to believe that meaningful classroom assessment which reveals students' understanding and levels of development in understanding a mathematical idea will result in, or at least be associated with, higher levels of student achievement [Richardson]. Teachers who allow students to internalize concepts and adjust teaching based on what they know about their students' development of concepts seem likely to produce students with more confidence in their ability. This confidence should translate to higher levels of achievement.

Instructional decisions that are made based on (or connected to) classroom assessments may result in improved achievement of students if those decisions reflect the following:

- support students in making connections
- move students to the next level of thinking
- build on strengths of students
- attend to individual students' needs and account for specific student needs
- do not make the curriculum "stagnant" in the sense of staying on a particular topic until all students "get it"
- allow the teacher to identify roadblocks for students and remove them

Although there is little research that connects staff development with student outcomes, there seems to be even less literature that links classroom assessment with student outcomes. It is unclear whether it is even reasonable to expect these links to be documented. Maybe all that can be said is that student outcomes are associated with, rather than closely linked to, effective classroom assessment techniques.

Classroom assessment should provide information to help teachers know when to push students to move to the next level of thinking. Assessment should help the teacher plan instruction that builds on the strengths of students [Bright; Parker]. We have long known about the "teachable moment" but now we need to think about what might be an "assessable moment" in the classroom. We do not know enough about how teachers decide when to assess and what to assess.

The type of feedback teachers give may be important tools for helping students develop the needed skills to set targets and measure progress toward goals. Often, teachers make comments such as, "You did a great job." Students might be better served by comments such as, "Let me tell you what I like about your thinking." or "Tell me how you got that answer?" or "When does your strategy work and when won't it work?" (Judith Sowder's paper includes a summary of some suggestions about effective and ineffective feedback.)

One point that surfaced in the discussions was that teachers may want to consider giving feedback on randomly selected problems rather than on every problem that a student solves. This would relieve some of the burden on teachers without letting students opt out of being engaged with all of the problems (since students would not know ahead of time which problems were to be given feedback).

No matter what a teacher does, she cannot understand mathematics for her students. It is ultimately the responsibility of students to understand. Traditional instruction seems to be based on the notion that "if you do what I do, you will understand what I understand." This is clearly a faulty assumption. Performance and understanding are not equivalent. When traditional instruction and traditional assessment are used, students soon learn how to play the game of giving the teacher what she wants rather than struggling to understand the content. Engaging students more deeply in the assessment process may help to wean students away from playing these "games." Discussion of ideas becomes one of the key ways to keep students engaged in self-assessing.

Teachers need to devote some instructional time to helping students acquire the skills necessary to set targets and to make these self-assessments [Joyner; Lindquist; Midgett; J. Williamson]. Responsibility of students for setting learning goals and assessing their own thinking (and their progress toward their goals) begins at primary level and continues to grow through high school (and beyond). There is some anecdotal evidence that, at least at high school level, students set learning targets for themselves that are too low. Teachers need to work with students to set high targets that both students and teachers are comfortable with. Students need to value understanding if they are to be expected to put forth the effort to do good work on assessment tasks. Students need to believe that mathematics should make personal sense.

Students also need to be educated about what the assessment activities are designed to reveal about their thinking. Students may need to be taught how to carry out similar assessments of their own work so that they will not be "surprised" at what the teacher says about that work. Students may also find it profitable at times to score each other's work. Students may need to be asked to defend their work, first to the teacher or classmates who already know the goals and second to a "classroom outsider" (e.g., parent) who doesn't know the learning goals. All of these ideas might help students validate for themselves what the teacher says about their performance. In turn, students will learn whether what the teacher says about their performance is helpful in improving their understanding.

Information generated by classroom assessment needs to relate to students' grades and to the schools' and teachers' goals for learning [Joyner]. But clearly not all assessment activities need to be graded. Communicating the results of classroom assessments is important.

Teachers' Knowledge

Classroom assessment is influenced by teachers' beliefs and attitudes [Bush; Joyner]. Knowledge alone is not a good predictor of how teachers will carry out assessments or what they will do with the information they get, since the information they get is itself influenced by beliefs and attitudes. Although this may not strictly qualify as something teachers should "know," it is important to acknowledge the role of beliefs and attitudes in the mix of what teachers bring to the task of classroom assessment.

Teachers need to have extensive mathematics knowledge (both conceptual and procedural) in order to implement effective classroom assessment [Barnett; Parker]. A common belief was that teachers cannot assess what they do not know. Sitting beside this belief was the companion belief that teachers can improve their mathematics knowledge during the process of interacting with students. However, it *does* seem true that teachers need to understand how major mathematics concepts develop in complexity over time in order to assess students' thinking. Knowledge acts as a filter for what teachers hear from students, so if teachers' knowledge is faulty or severely limited, they seem likely to misunderstand what their students say and do. Perhaps it is more accurate to say that teachers have difficulty interpreting responses that do not fit within what they know about mathematics.

Part of understanding mathematics is understanding that learning is a sense-making process that students need to experience for themselves. It is a developmental process that is difficult, if not impossible, to place into the context of achieving "mastery" of topics. Teachers need to know that acquiring understanding of the "big ideas" of mathematics is often a long and involved process.

It was generally conceded that few teachers or administrators have had adequate opportunities to develop mathematical understanding at this deep level. Teachers and administrators are often under-prepared to recognize where students are in their development of mathematical ideas. Too, there has been a shift in expectations of what constitutes appropriate mathematical behavior for students. This shift has clear implications for what instruction should look like and consequently what knowledge teachers need to have in order to carry out this changed instruction and the concomitant changed classroom assessments. Without this knowledge teachers may be unable to carry out effective classroom assessment. Unfortunately, the current expectations for students may be radically different from the mathematical experience that teachers had when they were students. The visions that teachers have of "mathematics instruction" based on their own experiences may not be adequate to support them in creating the mathematical experiences they are being asked to create for today's students.

In addition to mathematics knowledge, teachers need detailed understanding of the curriculum [Joyner; Moss; Richardson; Webb]. If teachers do not understand the "intended" curriculum, they may not implement it; they may implement only what they feel comfortable with. Teachers need to know how their "implemented" curriculum differs from the intended curriculum set by local curriculum standards and how their assessment of the implemented curriculum may not adequately address students' learning of the intended curriculum.

Teachers need to know how to respond to incorrect or partially correct responses from students [Berenson, Vidakovic, & Carter; Gunter; Jackson; Webb; Zawojewski & Silver]. Teachers need a repertoire of prompts to use to probe students' thinking. Some can be generic, but many have to be created on the spot, and choosing appropriate prompts often requires strong mathematics knowledge. It is not clear whether teachers can learn this "on their own" or whether acquiring this knowledge requires seeing such prompts in action. Video of effective probing of students' thinking might serve an important role in this process, though interacting with an inservice leader might prove to be a more effective way for teachers to acquire this knowledge. No matter what way is chosen, however, it may be the discussion (i.e., debriefing) around questioning that is most important for developing teachers' knowledge.

Once information is gathered and interpreted about students' thinking, teachers need to know what to do with that information. As was pointed out earlier, the focus of the conference was on using this information to inform instruction. In the past, there was considerable emphasis on analysis of "error patterns" that might emerge from watching students' performance. In the current reform movement, there appears to be more emphasis on identifying students' levels of thinking, related either to conceptual development or solution strategies (or both). Along with helping students to understand their errors, teachers are attempting to move students to higher levels of thinking. Perhaps underlying this shift is the belief that teachers cannot expect that errors will "disappear;" rather, teachers need to help students understand their thinking and how what they do is related to the underlying mathematical ideas.

Conference participants suggested many other things that they thought teachers should know:

- How can teachers help students succeed on end-of-year tests?
- How can classrooms be connected with end-of-year tests?
- What are the targets and goals, and what are the multiple paths that students might take to reach those goals?

- How can teachers follow the development of students' thinking through the year?
- How can teachers really "listen" to students to know what they understand?
- How can teachers question effectively?
- How do students think about mathematics?
- What information about students' thinking is most important?
- What are some benchmark behaviors that indicate the development of students' mathematics knowledge?
- How do students know when students exhibit mathematical understanding?
- How can various tools (e.g., manipulatives, technology) reveal differences in students' thinking
- What are the support systems available to teachers (e.g., from school administration, the school district, the state department, peer discussion groups)?

Teachers need common, clear mathematical language for use in the classroom. First, they need precise language to use with students. Elementary teachers, for example, often work very hard to help students develop (i.e., socially negotiate within the classroom) common understanding of important words (e.g., fair sharing). By middle school, however, teachers often implicitly assume that all students have already developed common understanding of words (e.g., operation). All teachers need to recognize the ambiguity of words and make sure that common, mathematically correct meanings are made explicit. Second, teachers need clear language to communicate to students and others outside the classroom what students know. It seems possible that the vague language that students use in talking about their own knowledge may actually interfere with the teacher's ability to diagnose students' thinking and then plan instruction. When vague language is used by teachers it may reflect their own lack of clarity about mathematics or assessment or both.

Teachers need to know that classroom assessment is a continuous process of gathering information about student understandings and making inferences based on that information [Webb]. It is a task that is on-going.

Teachers have not been nurtured in criteria about knowing how to determine what students know. Teachers need more knowledge of students' mathematical development [Berenson, Vidakovic, & Carter].

Similarly, inservice leaders have not been nurtured in criteria about how teachers develop in their implementation of classroom assessment. We need benchmarks of teachers' development. Teachers need to know how they think about their own thinking, about mathematics, about instruction, and about students' development. Thinking about their own thinking may have to precede teachers' thinking about students' thinking.

One key attitude is whether a teacher is teaching mathematics for its own sake or mathematics as a tool. These different perspectives certainly influence choices of tasks, classroom environment, deciding what is important to assess, and so on. Current reform suggests that a balance between the two is important, but individual teachers will almost certainly choose to emphasize one perspective more than the other, though the choice may be implicit.

Another key cluster of beliefs is those beliefs about what mathematics students are capable of learning [Berenson, Vidakovic, & Carter]. There is much talk in the reform movement of "mathematics learning for *all* students," but there are clearly many different variations on this theme. Teachers need to be explicitly aware of what their views are on this goal. If teachers assess the match between students' knowledge and their expectations of where they should be at a particular age/grade (rather than what students are actually thinking about mathematics), there will tend to be caps set on what students can learn. Every set of curriculum guidelines specifies what should (and therefore, what should not) be taught in each grade level. These specifications sometimes set limits on what students are given opportunities to learn at each grade. Students can often learn more than curriculum guidelines expect them to learn.

Supporting Teachers

Perhaps the most important characteristic of support for teachers is that it should be sensitive to the needs and responsibilities of teachers [Lindquist]. Whatever is suggested needs to be both practical and effective at helping teachers understand, at a deep level, the thinking of their students.

The obvious technique for supporting teachers is the development of effective professional development that promotes good classroom assessment practices [Briars]. One of the side benefits of such inservice seems to be professional revitalization of teachers who participate. By empowering teachers to understand their students' thinking, teachers find much more intrinsic satisfaction with their jobs. Teachers who can truly effect significant changes in students' thinking seem much less likely to experience burn-out and leave the profession. There are many examples of this kind of inservice (e.g., CGI, QUASAR), but there may be a need to acknowledge those examples more publicly.

Developing effective inservice must be preceded by the development of frameworks of mathematical knowledge (e.g., van Hiele levels of geometric thinking) and of students' thinking that will provide the contexts for teachers to use to interpret what students say and do. Along with these frameworks, techniques (e.g., problems to use, questions to ask) need to be developed for teachers to use to generate important information about students' thinking. Hopefully, these frameworks will provide mileposts that help teachers know when students are beginning to understand important mathematical ideas. One way to begin to develop some frameworks might be to study students' responses to standardized test items (such as the NAEP items). Seeing patterns across many students' responses might help interpret students' responses in any particular classroom. (Kenney & Silver, 1997, provides more information on the NAEP items.)

In a related way, staff developers need frameworks for understanding teachers' development and techniques to use to generate important information about teachers' thinking. In part, using these frameworks will allow staff developers to model classroom assessment techniques, but more importantly, these frameworks will assist in the planning of more effective staff development.

It is important to remember, however, that frameworks are lenses for understanding thinking. They are *not* value free; any framework represents particular views on students' development, teachers' development, and mathematics. This notion was captured during the conference in a diagram presented by Chuck Thompson (Figure 1) during one of the "reporting back" periods.

This diagram captures the notion that teacher development proceeds in stages. At each successive stage, teachers become more expert, so they can capture more of the evidence of students' thinking and distill that evidence coherently. This diagram helped conference participants make explicit a number of important questions. What does it mean to "look at a student's understanding" of a mathematics concept? What are we looking for? How do a teacher's beliefs and knowledge affect the interpretive lens? What is seen and what is missed at each stage? How does the information influence instructional decisions?

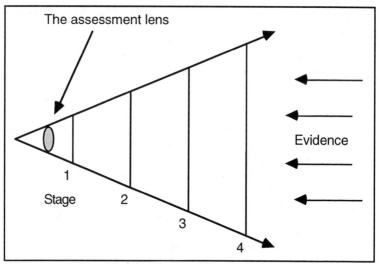

Figure 1.   A Lens Model of Stages of Teacher Development

Teachers should be supported by opportunities to reflect with other teachers on their growing knowledge of students and mathematics [Bright; Webb]. Time must be set aside for reflection. Hopefully schools will begin to build more of this time into their schedules, but teachers must also be encouraged to set aside personal time for reflection. It is probably true that training in, and implementation of, classroom assessment along with personal reflection on students' thinking combine to result in better instruction.

Video might prove to be a powerful tool for helping teachers become better at assessing. Cases may also be powerful [Barnett]. The discussion of, and reflection on, video or cases may prove to be, however, the most important aspect of this strategy. Teachers need to become better at considering alternate interpretations of what they hear students say and see students do.

One perspective on reflection is that teachers should become "action researchers" in their own classrooms. Considerable support might be afforded to teachers if we could help them to take on this type of role and to begin thinking of themselves as researchers. This could serve as a vehicle for collaboration between K-12 and higher education faculty to move classroom assessment into the "spotlight."

Teachers need to be "mentored" in developing personal understanding of mathematics so they begin to see what the role of a "facilitator" is. Then they can hopefully learn to do some of these things with their students.

Universities may want to develop courses on classroom assessment, perhaps with external grant support. The focus would clearly be different than standard "tests and measurements" courses.

Reflection may also be improved if there is "vertical dialogue" among all players (e.g., students, parents, administrators, teachers, teacher educators) about classroom assessment and students' thinking. Each group may benefit from interactions with the other groups. Teachers may need help in dealing with interpretation of assessment information for parents, especially for "higher level" mathematics (e.g., algebra) that parents might not be as familiar with.

Teachers need access to "good" literature on classroom assessment [Kenney]. But they may not have adequate time to do the searching that is required to locate this literature. Just surfing the net, for example, is not sufficient, since much of the information on the net may be inconsistent with quality classroom assessment.

Rich classroom assessment techniques should be embedded in curriculum materials. Teachers need models of ways to assess, and curriculum materials seem to be one vehicle for sharing these models. The task would then be to find out how to help teachers make sense of these models and make sense of, and use effectively, the information generated by these models. It is not clear what teachers can learn about assessment from seeing such examples. Without a strong knowledge base, teachers may not see the usefulness of either the models or the information they generate about students' thinking.

Teachers may need help in distinguishing between "good" and "bad" assessment tasks [Fennell; Richardson; Zawojewski & Silver]. By analogy, there are lots of instructional activities that students enjoy doing but that contain little mathematics or little chance to develop mathematical understanding. There are similarly lots of assessment activities that students (or teachers) enjoy doing but that reveal little about students' thinking. Good teaching is more than asking students to do interesting tasks, and "fun activities" may miss the mark with respect to assessment. Good lessons should help students learn important mathematics and should have embedded in them good assessment tasks that help the teacher understand students' thinking.

Not all teachers will find the same assessment techniques equally useful. Staff developers need to acknowledge the variability among teachers as well as the variability among students. Teachers themselves need to acknowledge the variability among their peers.

As with any change process, the people involved in making the change (in this case, teachers) may need to be confronted with the fact that they may have some personal inadequacies in teaching and assessment. But this must be done in a way that values the need for change rather than appearing to assign blame.

One of the illusions of staff development is that if workshop leaders give teachers tools (e.g., alternative assessment) they will use them. It is important to recognize, however, that it takes considerable knowledge to use sophisticated tools in meaningful ways. According to the *Assessment Standards* (NCTM, 1995), classroom assessment is difficult and complex. It is difficult to learn, difficult to do properly, and difficult to communicate the results clearly. It is probably important for teachers to understand these complexities and accept them as a fact of life, but to move on in spite of those complexities.

Areas of Special Concern

Technology
Little seems to be known about how technology influences the types of questions teachers need to use to probe students' thinking. It seems possible that students will learn differently (or perhaps learn different mathematics), but it is not clear how teachers might come to understand those differences [Clements].

It seems likely that technology will have to be embedded into curriculum and teaching before it will play a critical role in assessment. We cannot "add on" technology to assessment practices without first helping students learn how to use technology to learn mathematics.

Technology can help teachers and students record thinking. For example, hand-held technology might help teachers gather data in real time. Technology can store collections of evidence (e.g., portfolios), exemplars of various kinds of thinking, alternative rubrics for scoring, banks of tasks/problems, etc. for both teachers and students [Berenson, Vidakovic, & Carter; Sowder; Thompson; Webb].

In the not-too-distant future, there should be more technology that acts like an intelligent being, for example, in providing tutoring help [Clements]. Currently there are prototypes (algebra tutor, tutors for learning how to program) being developed. Most of these examples seem to be at the "upper" end of mathematics and require "upper-end" technology. Care needs to be taken in figuring out how this technology will interact with instruction and classroom assessment.

Equity
In terms of classroom assessment, the over-riding consideration is that it is important for teachers to know as much as possible about what each student is thinking. By overlooking any student (or any group of students) for any reason, the teacher would not be treating students equitably.

Rich instructional and assessment experiences in the classroom can help to "make up for" the lack of rich, out-of-school experiences by some students. Rich assessment strategies tied to rich instruction seems to be a powerful combination for all students.

Feedback that is unrelated to the content and expectations for performance can be a context within which discrimination will occur. Teachers (especially in elementary school) often comment on neatness or say things like, "I like the way that Chris is sitting." Without intending to do so, they may target these comments unevenly across students or send the message to students that aspects of behavior will substitute for understanding the mathematics. It seems more helpful for supporting students if the feedback is specifically related to the mathematics or the thinking involved in a solution, for example, by saying, "When you organized the numbers in a chart, I began to see what you were thinking."

Access to technology is also an equity issue. We have to be sure that all students have access to the same kinds of technology used in similar ways. Use of technology may widen the gap between students who perform well and students who perform poorly. Effective use of technology may require deeper understanding of mathematics, and so it may amplify the differences between students who know and students who don't know the mathematics. This is, of course, *not* an argument against the use of technology, but rather it is an argument in support of teachers' developing detailed knowledge of what their students know.

To re-enfranchise the students who "drop out" of mathematics (either actually or only through daydreaming) teachers need to understand what the students understand about mathematics. The goal should be to help them figure out mathematics and not to make them think like we do. As a partial preventative for having students drop out of a mathematics lesson, students might be asked to determine what is "just right" or "too easy" or "challenging" for them. This may empower students to take responsibility and may also reveal something about their level of mathematical thinking. Teachers should be encouraged to think about students as individuals, not just as adults packaged in younger bodies.

During the conference, several examples were offered about equity concerns. Disaggregating assessment data may be important because it will allow a truer picture of student performance to emerge, especially as related to equity. It may also provide information for reflection on the effects of instructional practices on different kinds of students. Disaggregating data allows teachers to see ways in which they might not have been meeting the needs of specific groups of students.

Preservice Education and Induction Years

Very little attention seems to have been given to preparing preservice teachers to use effective classroom assessment practices in their teaching. It seems possible that practices that are useful for novice teachers may be different from (or only a very restricted subset of) the practices that are useful for experienced teachers. Beginning teachers need to learn how assessment can inform their practice and help them become competent teachers more quickly [Jackson].

Perhaps, beginning teachers should teach only half a day and spend the other half of the day observing, reflecting on what they have learned, and so on. The move to ground so much of teacher education in actual school practices (in part, at least, by expecting preservice teachers to spend lots of time in schools) may be counter-productive, in that it succeeds in creating new teachers that are limited by the knowledge and practices of current teachers.

In addition, there needs to be faculty development for teacher educators on what classroom assessment is and how preservice teachers should learn about it. Teacher educators need to practice good classroom assessment in their own teaching (e.g., methods courses) so that preservice teachers have model from their own experience.

Needed Research

There were numerous comments made that could be construed as either areas for research or particular techniques for conducting classroom assessment research. There was not time during the conference either to elaborate on these or to prioritize them. Many of the working papers also identify areas of needed research [Barnett; Bush; Webb].

What do teachers think is important to know about students' thinking? Teachers will likely assess only what they think is important to know, but there is little information about those perceptions. How do teachers' perceptions change as they become more expert about classroom assessment or as they learn more about mathematics reform?

Are there patterns in teachers' thinking about the nature of classroom assessment? Are there stages of development in teachers' expertise about classroom assessment? Answers to these questions would be very useful to staff developers in knowing how to structure professional development experiences for teachers.

What catalysts help to maintain teachers' progress toward continual development of expertise about classroom assessment? Most conference participants commented on the need for support for teachers, but there was not always clear agreement on what support would be most useful for helping teachers remain engaged in reform of their classroom assessment practices.

It would be interesting to interview students with different backgrounds to find out their perceptions of the following questions:

- What kinds of feedback are helpful?
- What kinds of assessment questions are helpful?
- How does the language used by students or teachers impact students' responses to questions?
- Why do some students perform poorly on classroom assessment tasks?

- What do teachers do that encourages or supports learning?
- What do teachers do that impedes students' learning?

It would similarly be interesting to interview teachers with different backgrounds to find out their perceptions of the same questions. In comparing these two sets of interviews, do teachers and students identify the same things as important? How would different kinds of classrooms (e.g., traditional instruction, problem solving instruction) influence what students and teachers say about assessment? Context (e.g., classroom environment) seems likely to be an important influence on what is assessed, how it is assessed, how the information is interpreted, how the inferences are used, and so on.

Are classroom assessment practices similar across countries? How do assessment practices seem to impact the learning of students in various countries? Recent reports about the Third International Mathematics and Science Study (TIMSS) have re-activated interest in international comparisons of instructional practices.

There may be several areas of mathematical thinking that are "ripe" for research that will produce empirical evidence of the link between assessment and student achievement: for example, functions, statistics, and geometry. What are the theoretical mechanisms that support such linkages? How can those linkages be communicated in useful ways for teachers? More fundamentally, how do students learn the big ideas in mathematics (e.g., function, fractions)? How does self-assessment help students organize their knowledge of these big ideas?

Is it more effective (for learning, for attitude, or for development of self-assessment skills) to have students apply already created rubrics or to create their own rubrics? Does the answer to this question change as students become more used to using rubrics? What is the role of modeling by the teacher on how to use a rubric?

It was generally agreed that teachers really want to help students move from their current level of mathematical thinking to some higher level of thinking. Teachers can assist this process through questioning, posing an appropriate next problem, grouping students with others from whom they profit, and so on. Decisions about what question or problem to pose should be based on knowledge of students' thinking. But not all knowledge gathered is equally important and not all of it needs to be acted on immediately. We need more study of how teachers develop the expertise to make these decisions; that is, how teachers reflect on information and make inferences from that information. This process may be informed by what we know about how novices and experts organize their knowledge of a discipline.

Concluding Remarks

The range of comments during the conference was very broad. Participants brought diversity of expertise and experiences to the discussions. Their comments clearly reinforce the views that classroom assessment is quite complex and use of classroom assessment is critical for making sound instructional decisions and reporting what students actually know and can do. The environment within which teachers are expected to implement classroom assessment is also quite complex. There are many factors that influence the effects of classroom assessment on instruction, and ultimately on learning.

The discussions provided much "food for thought" about so many issues that it is neither possible nor desirable to over-simplify the situation by reducing the discussions to a list of "bulleted" conclusions. The primary benefit of the discussions seems to have been a sharpening of participants' thinking and the beginnings of a sharpening of the issues that might need to be addressed first. Much more effort is needed, however, to clarify these issues.

References

Kenney, P. A., & Silver, E. A. (1997). *Results from the sixth mathematics assessment of the National Assessment of Educational Progress*. Reston, VA: National Council of Teachers of Mathematics.

National Council of Teachers of Mathematics. (1995). *Assessment standards for school mathematics*. Reston, VA: Author.

Recommendations and Starting Points

Jeane M. Joyner
North Carolina Department of Public Instruction

George W. Bright
The University of North Carolina at Greensboro

In the ideal world, all students would be successful in achieving the mathematical goals of each grade or course. They would be confident in themselves as learners and would continue to explore new ideas and grow intellectually far beyond their formal schooling. In this utopia all teachers would be scholars and master teachers. They would be innovative and know well the mathematics that they are called upon to teach. They would understand how the mathematics of their classroom fits into the "big picture." In this ideal world teachers would have clear learning targets, be able to communicate their expectations to students and parents, and know how to assess those targets. They would be experts at guiding students to reveal their thinking and would then make appropriate inferences and instructional decisions.

During the working conference there were numerous conversations that involved the issues that are fundamental to creating classrooms where all students experience high levels of success in mathematics. From reading the summary chapter it should be clear that most of the issues are not new: the need for teachers to have a greater, deeper knowledge base in mathematics; research to give guidance for decisions that are made within the classroom and by policy makers; greater support for teachers; the need for classroom tested exemplars; and so on. The list is long, and it is similar to issues that are likely to be raised by any gathering of educators. All are important and complex; most are not likely to be resolved in the near future.

However, from the focused discussions we have interpreted recommendations and starting points. Because there was not sufficient time at the conference to reach consensus, these recommendations and starting points may not reflect the positions of any individual participant. Not all issues raised in the working conference will have associated recommendations, since we do not know how to resolve all of the issues. Rather, we have tried to cluster ideas that appear to belong to major issues and to craft specific recommendations that can be implemented now rather than waiting for the ideal world to arrive.

Summary of Issues Addressed in Discussions

Classroom assessment involves the gathering of information by teachers for the purposes of informing instruction and monitoring students' progress. Classroom assessment is not intended to rank students against each other but rather to give insight into the thinking, reasoning, and understanding of students and, at times, to measure their progress against standards. Other kinds of assessments have the purpose of ranking students, and it is important not to confuse either the kinds or purposes of assessment. Teachers should be allowed and encouraged to use classroom assessment for the purpose of improving instruction, independent of whether external assessments are also required for other purposes.

It seems counterproductive to the overall emphasis on greater student achievement to put in place external assessments that are so "high stakes" that they encourage people (e.g., teachers, administrators) to focus only on multiple-choice assessments as the primary assessment tool while at the same time trying to align mathematics instruction with current reforms. It may be that teachers and administrators are unintentionally allowing external assessment to drive instruction in the "wrong" direction, in the sense that teachers work too hard to make their classroom assessment "match" the form and focus of those external assessments. It seems plausible that if students are learning mathematics deeply, then scores on external assessments will reflect that knowledge. Teachers do, however, need to be aware of the nature of any external assessments that their students must complete.

Linking Instruction and Assessment
Classroom assessment and instruction should be closely linked. Planning for both instruction and assessment should build on identification of clear learning targets. Information from classroom assessments should help teachers decide how to help students expand their thinking. More knowledge is needed, however, about how teachers decide what information to gather and how to use that information in their decision making.

It is unclear whether teachers at all grade levels should
to learn equal amounts about their students individuall
kinds of interactions with students across the grades might influence
what teachers can learn about students. However, classroom assessment
should be equitably planned and implemented. Teachers should gather
information about all of their students and then make instructional
decisions that will assist all students to learn more mathematics. Too,
there should be more "vertical dialogue" among all stakeholders across
grade levels interested in classroom assessment. Interacting with other
groups can assist the process of reflection by allowing a variety of
inputs for that reflection.

Because teachers are assessing individual students' understanding,
classroom assessment by its nature is quite variable across classrooms.
This potentially creates some tensions among teachers in the sense that
they will be doing different things, and yet all teachers will call their
actions "classroom assessment."

Improving classroom assessment seems to be a task that any
teacher or building faculty could take on as a project with relatively
little initial cost, even though the "emotional" cost involved in making
changes might be somewhat higher. The potential payoff seems great,
so the investment seems important. The core of assessment must
remain focused on important mathematical ideas. Educating everyone
to implement this kind of classroom assessment may take some time.

Teachers and Students Working Together

The main job of students is to develop understanding of
mathematical ideas, and the main job of teachers is to help students do
their job. Teachers need to have considerable skill at monitoring
students and making decisions. Teachers must envision clear learning
targets and have a well-developed sense of what constitutes good (or at
least, acceptable) work from students. They must gather a wide range
of evidence of students' thinking, employ many perspectives to build
understanding of students' thinking, and know how to make inferences
from those data. Teachers should avoid over-generalizing what students
know, but rather, take a somewhat conservative approach to interpreting
what students say and do.

In the classroom, teachers must be able to engage students in
meaningful dialogue (both with the teacher and with other students)
about important mathematics. They must create an atmosphere in
which it is acceptable (and perhaps, even desirable) for students to

expose their thinking to public scrutiny. Teachers must listen carefully to what students say and to make sense, frequently on the spot, of what they hear.

Feedback from teachers to students seems to be a critical influence on what students learn. Feedback should help students focus on understanding of the content rather than on just obtaining correct answers. Having benchmark behaviors about student performance would help teachers, staff developers, and students all do better work. There needs to be agreement about what should be viewed as indicators of understanding.

Students need to have multiple avenues for demonstrating their learning. Students need encouragement to assume responsibility for knowing as well as setting high learning targets, for self-assessing their progress toward those targets, and for working toward mathematical understanding. A goal is for students to develop confidence and perseverance so that they will engage in assessment (and, of course, instructional) tasks to completion. To the extent possible, teachers should involve students directly in the assessment-design process. Classroom assessment should not artificially restrict what students can demonstrate about their understanding or how they should approach problem solving tasks.

Well-prepared Professionals

Teachers need considerable background knowledge in order to carry out quality classroom assessment. They need to understand mathematics deeply, have a good sense of the curriculum they teach, and know how what they teach fits into broader curriculum goals. They need to understand the ways that mathematical ideas develop in students' minds and the kinds of strategies that novices bring to the task of mastering content information. They need to know how to gather information from students and then to make inferences from that information.

Information needs to be shared with teachers on the variety of assessment techniques that are available and the differences in the information generated by each type. Each kind of assessment carries somewhat different expectations for students in terms of how and what they communicate about their understanding. Teachers need to know how to integrate the potentially inconsistent messages that come from use of multiple types of assessment.

In addition, teachers need to be willing to reflect on what they have heard students say and what they have seen students do. Reflection should typically not be done in isolation. Teachers need to be able to share their reflections with colleagues in order to refine the inferences drawn from the evidence. Teachers need to be willing to change their instruction in response to these inferences. Teachers need to be willing and able to share their inferences with students, with students' families, with colleagues, and with others in the larger community.

Classroom assessment is an opportunity for teachers to extend their knowledge of mathematics, students, and teaching itself. If students respond to assessments in predictable ways, then teachers can be confident that their knowledge of mathematics and of students is accurate and robust. If students respond unpredictably, then teachers should decide whether they need to reorganize or extend their knowledge about their students. There are many other filters, however, that might affect both what a teacher assesses and how a teacher interprets the information generated during the assessment. Teachers' prior experiences as well as their perceptions of students' capabilities, the curriculum, and community expectations for students are obvious influences.

There may be stages in teachers' development of implementation of classroom assessment. Frameworks are needed to help identify these levels of development. Effective staff development programs and materials (e.g., books, videos) need to be developed to help teachers learn to implement quality staff development. Examples of classroom assessment, either as "stand alone" examples or as part of curriculum materials, would be very helpful. Because implementing better classroom assessment takes time, effort, and resources, teachers need to be supported in their working environments (e.g., through school scheduling practices) as they begin to make changes. Both inservice and preservice preparation of teachers and administrators needs to include opportunities for them to develop sufficiently deep mathematical knowledge so that they can implement quality classroom assessment. Inservice for teacher educators and staff developers may be as important as inservice for classroom teachers in the whole process of implementation of classroom assessment. Teacher educators need to know how better to assist preservice teachers to enter teaching with knowledge of classroom assessment and assist inservice teachers to change their existing practices.

Continuing Research

Many other aspects of classroom assessment need to be investigated systematically. How does technology influence the types of assessment used, the types of information generated, or the types of inferences made from that information? What do teachers think is important about students' thinking? What support systems help sustain teachers' interest in implementing quality classroom assessment? What are the practical and technical issues related to classroom assessment? What do students think about classroom assessment practices? How are students' and teachers' views similar or different? How does classroom context influence what classroom assessment is done or what effect it has? Internationally, are there similarities and differences in classroom assessment practices and how the information derived is used? How do students and teachers learn to develop and use rubrics?

Recommendations and Starting Points

One persistent theme in the discussions was providing support for teachers as they come to understand the potential power of classroom assessment and begin to integrate classroom assessment practices in their instruction. In fact, almost every issue listed earlier in this chapter relates to this very broad and pervasive theme.

A second persistent theme was the focus on helping teachers make inferences about students' understanding, not just their performance on tests. Tests, including externally developed tests such as the SAT, do provide limited information about students' understanding, but the focus of classroom assessment is helping teachers learn about their students' understanding on a day-by-day basis.

Linking Instruction and Assessment

Probably the most critical part of implementing classroom assessment is that teachers need to be clear about the learning targets and how teaching to and assessing these learning targets impacts student achievement. Without that, the data gathered about students' performance will be meaningless and the inferences drawn from the data are virtually certain to be invalid. In addition to clear learning targets, however, there needs to be agreement about what evidence is acceptable for showing attainment of these targets. Setting targets and deciding on acceptable evidence requires that teachers and others interested or impacted by the assessments have opportunities to determine what is most important to teach and to assess and how to go about assessing those things.

Teachers' knowing more mathematics is not enough for accomplishing these tasks. Teachers need to know how the mathematical ideas connect with each other and with real-world applications. They need to be able to describe what student performance will look like when students have learned the appropriate mathematics. Further, teachers need to know the assessment strategies that will generate information that will document attainment of the goals. It would also seem quite helpful if there were frameworks of the important mathematics that students are supposed to learn, frameworks of how these important ideas develop in students' minds, and frameworks of problem solving strategies that students employ when dealing with this content.

In short, teachers need to become more reflective about the classroom assessment process. Teachers should look at and question what they are doing and what students are learning and then decide whether changes are needed. They should clearly distinguish the purposes of their personal classroom assessment and the purposes of external assessments that might be imposed by those outside the classroom. But becoming more reflective may not be a prerequisite to beginning the integration of classroom assessment in instructional practice; it might proceed along with greater use of classroom assessment practices.

RECOMMENDATION 1.1: All those involved with education need to understand that classroom assessment is a process, not an event, and this process is linked to instruction.

Classroom assessment is a multi-dimensional process (c.f., NCTM, 1989, 1995), including setting learning goals and planning how to assess, gathering information from students, interpreting the information and making inferences about students' understanding and progress, and making decisions and taking action. Teachers can enter the process at any point, so long as they remember that they are working within a larger process.

STARTING POINTS
- *Encourage teachers to explore the classroom assessment process as a whole but to begin at the point that makes the most sense for their particular instructional setting.*

- *Inform policy makers, perhaps through specific assessment examples or examples from specific classrooms, about the role of classroom assessment in influencing quality instruction and learning.*
- *Ensure that teacher evaluation systems include criteria to determine the extent to which teachers are using classroom assessment to monitor students' progress and to plan instruction.*

RECOMMENDATION 1.2: Agreement should be reached, among teachers and others outside the classroom, that classroom assessment and external assessments are designed and used for distinct purposes and that each needs to be valued for its unique contributions to students' achievement.

What happens on a day-to-day basis should be driven by what teachers know they are responsible for teaching and what they know about their students' thinking. For example, classroom assessment and end-of-year assessments should be based on the same global learning goals, so anticipated end-of-year assessments should not dictate what happens daily in the classroom.

STARTING POINTS
- *Initiate conversations among educators about the various kinds of assessment and their uses, so that the different purposes are clear.*
- *Provide information to teachers, families, and others in the community about the different kinds of assessment and their purposes.*
- *Focus the use of external assessments on their intended purposes.*
- *Ensure that decisions about students are based on a balance of information from classroom assessment, end-of-year assessments, and other external assessments.*
- *Monitor the influence of classroom assessment and external assessments on individual and school-wide performance.*

RECOMMENDATION 1.3: Incentives should be developed for encouraging teachers to reflect more deeply on their instructional practice and to find ways to refine that practice.

It continues to be important to upgrade the importance of teachers' self-evaluation of teaching as a means of improving professional practice. Integrating classroom assessment and instruction seems like a reasonable strategy for making such improvements. In this era of accountability for all students' learning, teachers have a moral and legal responsibility to try to understand what and how students are thinking so that they can make appropriate classroom decisions that will enhance that learning. This responsibility goes far beyond just assigning grades.

STARTING POINTS
- *Establish mentoring opportunities within schools so that teachers might work together for improvement of instructional practice.*
- *Establish "listservs" along with and as follow-up activities for professional development projects so that teachers might find mutual support.*
- *Compensate teachers for advanced course work focused on reflection about instructional practice.*
- *Include teacher reflection as a central component of teacher development projects.*
- *Help principals and others who evaluate teachers to learn how to look for and encourage the use of classroom assessment information for modifying instruction to match the performance levels of students.*
- *Provide district recognition and rewards for teachers who refine their instructional practices based on serious reflection and self-evaluation.*

RECOMMENDATION 1.4: Frameworks should be identified or developed that organize critical areas of mathematics (e.g., fractions, functions) more coherently and that make students' thinking more comprehensible.

Examples of these frameworks are the addition/subtraction problem types developed for Cognitively Guided Instruction and the van Hiele levels of thinking in geometry. One of the uses of content frameworks is to help teachers clarify both learning targets and acceptable performance indicators of those targets. The frameworks would also help teachers make more valid inferences about what students know and can do. Frameworks appear to be extremely useful in helping teachers make sense of students' developing understanding.

STARTING POINTS
- *Fund projects, on a high priority basis, which will develop these frameworks and document through research both the accuracy of the frameworks and their effectiveness at helping teachers to understand students' thinking.*
- *Disseminate validated frameworks through funded projects, professional organizations, and professional journals.*
- *Assist at district levels and state levels the process of dissemination of frameworks through publications, professional development, and individual contacts.*

Teachers and Students Working Together

In the same way that reflection is important for teachers in their development of understanding of classroom assessment, reflection is important for students in the development of their understanding of mathematics. Indeed, the role of students in classroom assessment is clearly under-represented in the literature. Students need to play a bigger role in the design of classroom assessments, so that they have a good chance to demonstrate what they know. Students also need to assume some responsibility for completing assessments to the best of their abilities and for engaging in self-assessment. If students cannot demonstrate, or are not willing to reveal, what they know, then teachers' inferences may be inaccurate.

All parties (e.g., teachers, researchers, curriculum developers, test designers, legislators) need to have models of effective classroom assessment materials. These might include tasks, rubrics, questioning strategies, observation protocols, interview questions, and so on. These tools should be accompanied by frameworks for interpreting students' responses, guidelines for constructing new tools or modifying existing examples, suggestions for when to use the tools, and so on. The resulting "tool kit" should be accompanied by an inservice program that helps teachers learn to use the tools effectively and efficiently.

The models might be stand-alone examples that can be used on demand by teachers or they might be part of larger curriculum materials (e.g., NSF-funded mathematics instructional materials). When the models are embedded in instructional materials, it might be easier to communicate to teachers how student performance might influence future instruction, though in all cases, attention to this issue seems critical.

It seems impossible to create "value free" assessments. All assessments reflect choices about what is important to know and how that knowledge should be demonstrated or communicated. Models of classroom assessment should make clear the value base within which they were designed.

RECOMMENDATION 2.1: Strategies need to be developed for helping students become more involved in classroom assessment and in self-assessment.

This task might be taken on by teachers working in small groups, by university/school partnerships, by teachers and students working together, or in other ways. Strategies could be developed as part of a professional development program for teachers, as part of the development of curriculum materials, or through specially funded projects. Integrating the strategies throughout instruction might prove to be more difficult than developing the strategies themselves, since use of these strategies might need to be based on underlying values and assumptions that are not shared by all teachers.

STARTING POINTS
- *Involve students in creating and applying rubrics to evaluate mathematics performance.*
- *Provide opportunities for students to engage in self-assessment on a regular basis.*
- *Develop portfolios for which students and teachers together determine the work included.*
- *Initiate student-led conferences with their families as a means of providing feedback to families about students' performance and of giving students greater ownership and responsibility for their own learning.*
- *Develop techniques to help students become better at self-assessment.*

RECOMMENDATION 2.2: Field-tested examples of classroom assessments need to be developed and widely disseminated.

Different forms of assessments may need to be developed to account for different kinds of available technologies (e.g., calculators, computers, manipulatives). Models should clearly spell out the frameworks that underlie their development (including the relevant learning targets) and should help teachers know how to make sense of students' responses (perhaps through providing annotated samples of students' work).

STARTING POINTS
- *Collect and disseminate assessment examples that are specific to courses or grade-level expectations.*
- *Create assessment examples that are accompanied by clear and detailed discussion of how the information gathered can be interpreted to determine whether students have attained a particular learning target.*
- *Identify and demonstrate how different technologies might alter the ways that each kind of assessment might be used.*
- *Publish and widely disseminate pamphlets or videotapes which show how different assessment strategies can be used within a classroom.*

RECOMMENDATION 2.3: *Results of classroom assessments need to be used explicitly by teachers to modify instruction.*

The primary goal of classroom assessment is the improvement of instruction to increase student learning. Once examples of high quality classroom assessment are available, teachers need to be encouraged to use the information generated to modify their instruction. All educators need to train their eyes and minds to see what is important about students' learning and ignore what is not important. Classroom assessment is a convenient lens for beginning to observe students' understanding more clearly.

STARTING POINTS
- *Create opportunities for teachers to talk with each other about how to use assessment information to modify instruction; for example, through grade-level meetings, department meetings, professional development projects, etc.*
- *Publish cases in which teachers explain how they have modified instruction based on their knowledge of students' understanding.*
- *Counter the mind-set of "covering the curriculum" with "teaching for understanding."*

RECOMMENDATION 2.4: *Because teachers who use classroom assessment effectively have detailed information about individual students, teachers should play a central role in decisions about their students.*

The information that classroom teachers have about their students needs to be valued in any decision-making about those students. Teachers watch the development of students' understanding over many days, and they are in the best position to understand the effects of various decisions on their students. That expertise should not be subjugated to results of external assessments.

STARTING POINTS
* *Make decisions about students; for example, placement in special programs or courses; using multiple sources of evidence.*
* *Facilitate opportunities for cross-grade-level discussions about students.*

Well-prepared Professionals
There need to be many different ways for teachers to learn what they need to know about classroom assessment. Some will be formal programs, but others might be more informal approaches (e.g., electronic "study groups" similar to those developed through MathLine). All professional development programs, however, should help participant teachers develop common understandings of what their learning targets are and how they would know whether students have attained those targets. Particular professional development programs might focus, among other things, on how to adapt assessment tasks, how to question, what to look for in students' responses (e.g., what they need to probe), how to evaluate assessment tools to decide if those tools will provide needed information about students, and so on.

As with instruction for students, examples of high-quality professional development materials need to be developed for use with teachers. One possible product is a set of reader-friendly stories that exemplify various dilemmas of classroom assessment. The stories could facilitate discussion about how to begin, or continue to develop, effective classroom assessment practices. In these stories, teachers might find "soul mates" for the difficulties they are having individually (e.g., how to communicate with parents, how to find time to reflect) so that teachers would not feel so isolated. Other products might include videos, web sites, workshop manuals, and so on.

If classroom assessment practices are going to improve, teachers need time to learn more, to reflect, and to deal with the innumerable issues that will emerge. Teachers cannot be expected to do these things only on their own time; institutional support needs to be provided to assist them. Teachers need an institutional environment that encourages experimentation and provides support when experiments work or fail.

Institutional support needs to include time for teachers to do this important work and an attitude that encourages it. For example, schools might restructure the school day so that there is time for teachers to meet to talk about their efforts to integrate classroom assessment in instruction. This time might be daily or weekly planning time when teams of teachers could meet, formal professional development time provided through early release of students from class, time for new teachers (or teachers new to a grade level) to meet with mentor teachers to discuss classroom assessment issues, or time for teachers to visit each other to watch classroom assessment in action.

RECOMMENDATION 3.1: Develop effective professional development programs that help teachers understand and implement classroom assessment in a context that also deepens their knowledge of mathematics and pedagogy.

The goal is to help teachers acquire the knowledge they need to implement quality classroom assessment. Both inservice and preservice programs need to be developed, though what is emphasized in these programs might be different. Certainly, new teachers entering the profession need to be ready to implement classroom assessment effectively. Programs might be developed with grant support, by local districts, or by commercial publishers or professional development companies.

> *STARTING POINTS*
> - *Provide opportunities for teachers to learn about and practice classroom assessment, including effective classroom questioning strategies, as part of every professional development program that involves classroom instruction.*
> - *Schedule visits by the workshop leaders to participant teachers and visits among participant teachers (e.g., as follow-up to professional development programs) as a means of connecting their learning about classroom assessment to actual classroom practice.*

RECOMMENDATION 3.2: Develop materials to support effective staff development programs.

Just as teachers need materials to teach mathematics, staff developers need instructional materials to teach teachers effectively. These materials need to be designed to account for principles of adult learning.

STARTING POINTS
- *Create cases (either in print or videotape form) which illustrate ways that teachers use classroom assessment techniques to inform their instructional decisions.*
- *Develop print resources that teachers can use either in professional development projects or individually.*
- *Create materials that help teachers deal with the "tough issues" surrounding classroom assessment; for example, classroom management, management of information about students, asking good questions, etc.*

RECOMMENDATION 3.3: Restructure school days so that teachers can spend their time in ways that would support better classroom assessment.

All parties (e.g., teachers, administrators, legislators) need to be involved in these discussions so that the needs of all stakeholders are recognized and addressed. The support needed for teachers clearly goes beyond just professional development programs, so discussions should not be limited to "traditional" teacher-support strategies. For example, ways for developing an institutional "ethos" to support classroom assessment clearly need to be examined.

STARTING POINTS
- *Initiate discussions about options for restructuring school schedules.*
- *Provide common planning times for teachers to meet and discuss classroom assessment issues.*
- *Reschedule classes so that teachers can have quality discussions with individual or small groups of students so that teachers can better understand students' thinking.*

RECOMMENDATION 3.4: School budgets need to begin including alternative support mechanisms for teachers to develop, try out, and implement effective classroom assessment.

New technologies need to be used to improve classroom assessment, so funds need to be allocated to allow acquisition of these technologies. For example, the mathematics coordinator in a district might move some "staff development" money into a fund that would support the use of substitute teachers while regular classroom teachers visit each other to see classroom assessment in action.

STARTING POINTS

- *Support teachers to participate in conferences about classroom assessment with the expectation that they will share their new knowledge with colleagues.*
- *Develop sabbaticals for teachers so that they have time to learn how to conduct effective classroom assessment.*
- *Identify and compensate teachers who are particularly effective at using classroom assessment to act as mentors for teachers who are less expert.*
- *Provide substitutes for teachers so that they can visit other teachers to see classroom assessment in action.*
- *Provide release time to develop and modify classroom assessments and management systems.*
- *Designate funds for the initial purchase and continued maintenance of technology useful for implementing classroom assessment.*

Research

There is clearly much that is unknown about classroom assessment. To attempt to provide an exhaustive list would be self-defeating, but it is important to note that a focus on classroom assessment may shift the emphasis of research questions more toward concerns of learning and away from issues solely of teaching. But one of the most critical issues is how teachers' knowledge of their students develops. What are the major influences on this development? Is it mathematical knowledge? Is it knowledge of cognitive development? Is it attitude about mathematics or about students? Are there stages of development? It might be useful for researchers to tell stories of individual teacher's development that indicate the nature and importance of the influences, though other techniques for disseminating the results also need to be developed.

Evidence is needed about whether teaching in a problem solving, holistic thinking and reasoning way is associated with increases in student achievement as measured by different assessments (but most notably external, state-mandated assessments). Without this evidence it is difficult to prevent external assessments from driving classroom instruction in the "wrong" directions. Many accountability systems are based on an assumption that schools should prepare students to deal with complex problems as they might be encountered in the real world, yet state-mandated assessments are too often built around multiple-choice test items, which seem to be the antithesis of complex, real-world problems.

RECOMMENDATION 4.1: There needs to be more research that focuses on the issues related to quality classroom assessment and on expanding knowledge about ways that technology can support classroom assessment.

Some of this research could, and in fact should, be conducted along with professional development activities. But some might also need to be conducted independent of any particular program, with its accompanying value system. Funding for research might come from grants, from local schools, or from investments of time by researchers. Because the focus of classroom assessment is on helping teachers learn about their students' mathematical understanding, research on classroom assessment will need to address issues related to the development of mathematical understanding.

STARTING POINTS
- *Provide incentives for researchers to investigate the issues surrounding classroom assessment as it informs instructional practice.*
- *Provide incentives for researchers to investigate how technology can support classroom assessment.*
- *Provide opportunities for teachers to link with researchers in the study of classroom assessment.*
- *Disseminate research results to policy makers in ways that will inform the policies created.*
- *Disseminate research results to classroom practitioners.*
- *Disseminate research results to those who prepare teachers so that they will better model the links between instruction and classroom assessment in order that beginning teachers are better prepared to implement classroom assessment.*

Implementation
None of the recommendations can be implemented easily. Implementation will require considerable effort from many different people (e.g., teachers, school administrators, university faculty, policy makers, parents, students) at the local, state, and national levels. The difficulties of coordinating the efforts of all these groups are likely to be

great, but without that coordination, implementation efforts are likely to fall short. Even at a fairly local level (e.g., within a single school district), there are many players (teachers, students, parents, school boards) that need to work together. Each group has its own perspective on the issues, so making these different views explicit will be critical to success.

At the state level, legislative committees need to become actively involved in implementation of the recommendations. Mathematics educators may need to become more politically astute if they are going to be successful. In addition, university faculties can play an important role, both in preservice and inservice efforts. Coordinating these faculties will likely require that university administrators be invited and encouraged to participate in planning.

At the national level, professional organizations (e.g., National Council of Teachers of Mathematics, National Council of Supervisors of Mathematics, Mathematical Sciences Education Board) need to become actively involved in implementation efforts. The leadership provided by these groups will be critical or supplying a rationale for implementation efforts.

RECOMMENDATION 5.1: Responsibility for implementing quality classroom assessment must be shared by all stakeholders.

It is obvious for some recommendations and starting points where responsibility lies; for example, those who control funding must assume responsibility for implementing any recommendation that requires significant funding. For other recommendations and starting points, the responsibility is not so clear and in fact may need to be shared by different groups.

However, if people do not take responsibility, but rather assume that someone else will take responsibility, then nothing will happen. All stakeholders need to be willing to assume responsibility for things that they can effect. We hope that many people will willingly step forward and provide the leadership that will be necessary for implementation of the recommendations.

STARTING POINTS
- *Determine how responsibilities are to be shared.*
- *Create an accountability strategy for seeing that recommendations are wisely implemented.*

Concluding Remarks

Underlying all of the recommendations is the notion that it will generally be up to teachers to carry out the work of classroom assessment. For all of the recommendations to be implemented, however, the entire educational community needs to provide support for teachers in carrying out their work. All parties must keep in mind that the ultimate goal is the improvement of students' mathematical understanding. The intervening variable is the improvement of instructional practice in response to more detailed knowledge of day-to-day changes in students' understanding.

Hopefully, these recommendations will lead to development of initiatives at local, state, or national levels that will improve the quality of classroom assessment. We must move beyond talking about classroom assessment to become engaged in developing expertise in classroom assessment and in using classroom assessment to support instructional planning. Even small steps on this journey seem likely to produce noticeable benefits for students' learning.

Section 2

Position Papers

Implementing Standards-Based Classroom Assessment Practices

Diane J. Briars

Pittsburgh Public Schools

Standards-based assessment is an essential component of any standards-based mathematics program. Often, however, discussions of standards-based assessment focus only on high stakes assessments (e.g., district-wide norm-referenced tests, state mandated assessments) with little or no attention to classroom assessment. Yet, it is teacher-controlled classroom assessments, such as oral feedback in class, test results, grades, that have most direct impact students on a daily basis.

Classroom assessments give students regular feedback about their knowledge of mathematics and their ability to do mathematics, and thus influence their perceptions of themselves as mathematics learners. The mathematics that students have the opportunity to learn through the assignments they receive (enriched or remedial) and the courses they take (college preparatory vs. general math) is largely determined by grades and teacher recommendations, both of which are based on classroom assessment results. It is classroom assessment that provides teachers with the ongoing feedback used for instructional planning. Thus, although high stakes assessments can have a direct impact on teachers and administrators, influencing their choice of content and allocation of instructional time, it is ongoing classroom assessment that most directly influences students on a regular basis. As a result, it is essential that classroom assessment be of the highest quality, making the implementation of standards-based classroom assessment a key aspect of high quality mathematics instruction.

Over the six past years, I have been involved in supporting urban teachers in the implementation of standards-based classroom assessment in two contexts: (a) implementation of new elementary and middle school standards-based instructional materials in the Pittsburgh Public Schools, and (b) two professional development projects, CAM and

ACT, that focused on classroom assessment practices of middle school teachers. The Classroom Assessment in Mathematics (CAM) project supported teams of 4-6 middle school teachers in six urban school districts in "structured experimentation" with various forms of classroom assessment, such as, performance tasks, observation, journals, and projects. CAM teachers then served as leaders in the subsequent Assessment Communities of Teachers (ACT) project. In ACT, each district designed its own professional development plan to support middle school teachers in implementing standards-based mathematics assessment. CAM was supported by the Department of Education; ACT was supported by the National Science Foundation. This paper summaries some of the major lessons learned about implementing standards-based classroom assessment, and identifies promising entry points and supports teachers need to develop assessment practices that enhance students' learning.

The *Assessment Standards*

The *Assessment Standards for School Mathematics* (National Council of Teachers of Mathematics, 1995) has proven to be a valuable tool for working with teachers on classroom assessment. The *Standards* document defines assessment as "the process of gathering evidence about students' knowledge of, ability to use, and disposition towards mathematics and making inferences based on that evidence for a variety of purposes" (p. 3). Four broad purposes are described: monitoring student progress, making instructional decisions, evaluating student achievement, and evaluating programs. Six assessment standards provide criteria for judging the quality of assessments and assessment practices:

> *Mathematics*: Assessment should reflect mathematics that all students need to know and be able to do.
> *Learning*: Assessment should enhance mathematics learning.
> *Equity*: Assessment should promote equity.
> *Openness*: Assessment should be an open process.
> *Inferences*: Assessment should promote valid inferences about mathematics learning.
> *Coherence*: Assessment should be a coherent process.

The *Standards* describes shifts in assessment practices that result from the use of these standards. Together, the definition of assessment, the standards, and shift in practice were the focus of our work with teachers regarding their classroom assessment practices.

Assessment as a Process

Thinking about assessment as a process, rather than as an activity that is done after instruction is completed, is the basis of standards-based assessment. Defining assessment as the process of gathering evidence about student's thinking, that is, trying to determine "what's in students' heads" from visible clues (i.e., what students say, write, and do) seems very natural for most teachers. In fact, when asked to characterize the mathematical knowledge of a particular student and describe the evidence they used to formulate that characterization, most teachers cite a variety of sources of evidence: questions the student asked, verbal responses to teacher and student questions, observations of the student in action (e.g., solving problems, using manipulatives), and homework and test performance.

Most often, though, use of these multiple sources is informal and limited to assessment for the purpose of instructional planning or informally monitoring student progress. "Real assessment" is thought of as work that can be used to determine a grade, and as a result, is almost exclusively paper-and-pencil work. Consequently, we have found that various kinds of written assignments such as performance tasks, projects, and journals are good entry points for encouraging teachers, at least in grades 3 and beyond, to reflect upon and change their assessment practices. In attempting to use such assessments, three questions immediately arise: What are good tasks and where can I find them? How do I score (grade) student work? How do I use this work for a grade?

Design and Selection of Assessment Tasks

Fortunately, the answer to the question of where to find good assessment tasks is becoming easier as collections of tasks and prompts are becoming more available. New instructional materials (e.g., NSF-developed materials such as *Everyday Mathematics, Investigations in Number, Data, and Space, Connected Mathematics Project*) contain high quality assessments. Collections of tasks are also becoming increasingly available (e.g., Balanced Assessment materials; National Council of Supervisors of Mathematics (NCSM) task collection).

Selecting and designing tasks to assess conceptual understanding tends to be an overlooked aspect of classroom assessment. Typically, teachers use a limited range and type of assessment to assess conceptual knowledge. For example, to assess students' understanding of multiplication of whole numbers in elementary grades, many teachers simply ask students to solve multiplication equations (e.g., 4 x 5 =) or solve routine multiplication word problems. Rarely do teachers give tasks that provide direct evidence about students' understanding that multiplication describes situations involving groups with the same number of items in each group, for example, create a situation that could be described by 4 x 6 = 24, or explain what 3 x 4 means to someone else. Even more rare is the use of a set of different tasks designed to assess different representations or aspects of the same concept. Teachers clearly need a broader repertoire of techniques for assessing concepts.

In addition, to better assess conceptual understanding, teachers need more sophisticated understanding of how students learn particular concepts and what typical misconceptions might arise along the way. This need is an opportunity for linking research and practice by informing teachers about conceptual frameworks and tasks from research in particular areas, for example, the Van Hiele model for geometry or analysis of rational number concepts.

There is a cautionary note, however. In recent years, both the mathematics education community and the education community at large have been sending strong messages urging teachers to use "authentic" performance tasks. In our efforts to encourage teachers to assess mathematical problem solving, reasoning, and communication in addition to concepts and skills, we inadvertently may be giving the message that "bigger is better," that is, standards-based assessment means using complex performance tasks or portfolios. In short, we may have created the assessment analog to "using manipulatives is standards-based instruction." Too often, well-intentioned teachers are asking students to do long tasks and projects that require lots of time and effort, yet provide little evidence about students' mathematical knowledge other than their arithmetic skills. At the same time, short questions and tasks, such as those above, which provide valuable information about students' conceptual understanding are being rejected as "not authentic." Teachers need a more balanced message and more balanced sample collections of quality assessment tasks to help them become more effective assessors of students' knowledge.

Making Inferences from Student Work

The second question that arises about more open written tasks is how to score them. The perception is that performance tasks are extremely time consuming to score. We have found that need not be the case. The key is selecting a scoring rubric (scoring guide) with clearly defined performance categories that reflect students' understanding and use of mathematics.

R. Ready for Revision	**4+ Distinction**
The response accomplishes the purpose of the task or provides ample evidence that the performer has the mathematical power to do so. The execution may be faulty. An operational test for inclusion as an R is the following teacherly question: Can you give written feedback to the student that focuses his or her attention on the flaw that would be sufficient (without dialogue) for the student to revise the performance so it would meet the standard?	A distinguished performance is exciting -- a gem. It excels and merits nomination for distinction by meeting the standard for a 4 and containing special insights or powerful generalizations or eloquence or other exceptional qualities.
	4 Accomplishes the Task
	The response accomplishes the prompted purpose. The student's strategy and execution meet the content, thinking processes, and qualitative demands of the tasks. Even for tasks that are very open regarding content, the content chosen by the students must serve the purpose well. Communication is judged by its effectiveness, not by correctness or length. Although a 4 need not be perfect, any defects must be minor and very likely to be repaired by the student's own editing without benefit of a note from a reader.
	3 Ready for Needed Revision
	Evidence in the response convinces you that the student can revise the work to a 4 with the help of written feedback. The student does not nee a dialogue or additional teaching. Any overlooked issues, misleading assumptions, or errors in execution—to be addressed in the revision—do not subvert the scorer's confidence that the student's mathematical power is ample to accomplish the task.

Figure 1. New Standards Project Mathematics Task Rubric (Figure continued on next page)

M. More Instruction Needed	2 Partial Success
The response lacks adequate evidence of the learning and strategic tools that are needed to accomplish the task; or, the response has evidence of inadequate learning. Feedback to the student would not be enough. A teacher would have to interact with the student or teach more.	Part of the task is accomplished, but there is a lack of evidence—or evidence of lack—in some areas needed to accomplish the whole task. It is not clear that the student is ready to revise the work without a conversation or more teaching.
	1 Engaged in Task but with Little Success
	The response may have fragments of appropriate materials from the curriculum and may show effort to accomplish the task, but with little or no success. The task may be misconceived, or the approach may be incoherent, or the response might lack any correct results. Nonetheless, it is evident that the respondent tackled the task and put some mathematical knowledge and tools to work.
	0 No Response or Off Task
	When the response is blank, it is scored NR. When there are marks, words, or drawings unrelated to the task, it is scored OT. In either case, there is no evidence that the task was engaged.

Figure 1. New Standards Project Mathematics Task Rubric (concluded)

Pittsburgh teachers have found that the rubric (Figure 1) developed by the New Standards Project both provides valuable information about students' mathematical knowledge and is easy to use. The rubric is based on the key instructional question, "Does the response show evidence that the student knows the key mathematical ideas required for this task?" Responses which demonstrate adequate knowledge are considered to be "Ready for Revision" (R) responses; those that do not are categorized as "More Instruction Needed" (M) responses. R responses are further categorized as adequately completed the task (a *4* response), or as responses that, although flawed in some way (e.g., careless computational error, unclear explanation) nonetheless provide convincing evidence that the student could revise the response to be a 4 without additional instruction (a *3* response). In the M category, responses that show partial grasp of the mathematics in the task receive a score of *2*; those with little accomplishment, a *1*.

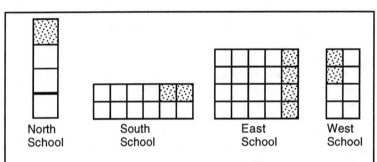

| North School | South School | East School | West School |

These are the plans for four different school playgrounds. The principals are discussing how they have set up separate sections for preschoolers and elementary school students. In each plan, the preschool section is shaded; the elementary student section is white.

South's principal says that South has the greatest fractional part of its playground for preschoolers. The other principals disagree. Please settle the argument and help each principal understand your answer. Use a diagram if it will help you explain which school has the greatest *fractional part* of its playground for preschoolers.

Figure 2.   Playgrounds Task

In a fifth-grade task (Figure 2) in which asks students are to label and compare fractional parts of regions, the rubric would be applied as follows:

4 The student compared the fractional parts correctly and gave a convincing explanation of the comparison. The comparison could be illustrated in a variety of ways, e.g., pictorially (e.g., circles showing 1/4, 1/5, and 1/6), using equivalent fractions with a common denominator, expressing fractions as decimals or percents.

3 The fractional parts are labeled and compared correctly, but the response contains a flaw, such as an explanation that is not very clear. The work suggests, however, that the student could revise this response to a 4 without further instruction in fractions.

2   The fractional parts are labeled correctly, but the comparison is incorrect.

1   The student is unable to consistently label fractional parts correctly.

In classroom use of this task, teachers would "score" the responses by sorting them into piles of *4*, *3*, *2* and *1* papers. The *4* and 3 papers suggest adequate understanding; the *2* and *1* papers suggest more instruction is needed. A closer look at the *2* and *1* papers will suggest the nature of the needed instruction. Detailed scoring and analysis of the nuances of each response are not necessary.

This rubric is very appealing to teachers because mathematical knowledge is the basis of the distinction between categories. Also, there are only four main score categories, with very clear distinctions between them, which makes the rubric easy to use. Pittsburgh teachers have also applied this basic rubric to larger tasks and projects, with extensions and elaborations when needed.

Making valid inferences about students' knowledge from assessment results is essential for assessment to enhance learning. Teachers need opportunities to reflect on the inferences they make, discuss them with others, and then test their inferences. Teachers can gain valuable insights into the inference process by looking at student work samples and asking a variety of questions: What inferences can be made from particular work samples? What other information would I like to have? Am I inferring too much?

There are two cautions to be kept in mind, however. First, rubrics may not represent the full range of mathematical understanding. Teachers need to be sure that all kinds of understanding can be captured by a rubric. Second, process has replaced product. Teachers still need to find a balance between product and process. Teachers need to examine both processes and products in order to make quality judgments.

Students' Role in Assessment

The NCTM Assessment Standards call for students to play an active role in the assessment process. This is the area in which we have seen the greatest change in teachers' assessment practices. In CAM and ACT, teachers engage students in activities designed to help students understand what will be assessed and what good work looks like. Students discuss and rate responses on performance tasks, develop their own rubrics, and have opportunities to revise their work to meet the criteria. Teachers report that such activities have had positive impact on student achievement.

Equity

In CAM and ACT, we have paid particular attention to equity issues in assessment, that is, providing each student the opportunity to show what s/he knows and is able to do, and also dis-aggregating data to monitor possible bias in assessment or instruction. This is the area that teachers have found the most challenging. Especially in middle schools, where teachers see 150 students per day and there is often limited flexibility in scheduling, teachers feel that are not doing enough to meet the needs of each student. Providing realistic models and strategies for promoting equitable assessment is a key issue for future work.

Closing Comment

Both CAM and ACT have illustrated that assessment *can* be a valuable catalyst for changing instructional practices as well as assessment practices. Assessment issues bring student thinking and learning to the forefront. As teachers focus more on student learning and get better information about students' mathematical knowledge, their instructional practices tend to become more student-centered.

Reference

National Council of Teachers of Mathematics. (1995). *Assessment standards for school mathematics.* Reston, VA: Author.

The Pro-Am of Assessment in the Elementary Classroom

Mari Muri
Connecticut State Department of Education

There is a Peanuts cartoon that shows two students seated in a classroom. Frame one: One student raises her hand and asks, "Today?! The test is Today? Yes, Ma'am, I'm surprised..." Frame two: The same student continues, "I thought maybe before the real test there'd be a pro-am." I will show how this golf analogy fits into the classroom as well as what happens on the golf course.

I live in a small community of about 11,000 people. Our town's claim to fame is the Tournament Players Club that hosts the annual Greater Hartford Open Golf Tournament. So, once a year, the town swells to almost ten times its size for over a week. Early in the week touring professional golfers, celebrity players, other amateur golfers, and thousands of spectators converge on our town for the pro-am tourney.

There are two reasons for this early attendance. One is to promote the golf tournament by having pros play with local amateurs and celebrities. These rounds get major television coverage, at least in the local area. The second reason is so that the professionals will get to know the course before competing for the big prize of the actual tournament. The latter reason is what I use as the analogy for my discussion about the power of classroom assessment as preparation for high stakes, externally imposed assessment.

Rationale

The professional golfers use the pro-am to get to know the golf course, try out a newly acquired club or putter, and check out pin placements and speed of the greens. No pro would go directly into the tournament with its big monetary award without practice on the course.

A comparison can be made between the practice before a tournament and the instructional activities and classroom assessments in which students engage: the pro-am before the "big" test.

The teacher's role is to provide opportunities for students to assimilate new learning and try new approaches to solving problems before being assessed for the "big prize." This means the teachers need to be familiar with the format and the content of the external test(s). As with driving and putting in golf, both content and format play major roles in testing. Two questions arise here:

- How does the teacher provide these practice opportunities for students?
- How does the teacher know when the students are ready for the test?

Providing Opportunities

The teacher must be familiar with all the tests the students will face. This is relatively easy with teacher developed and administered tests, but may be a challenge with externally imposed tests. To overcome this challenge, teachers need to avail themselves of test content specifications, item formats, sample items and performance criteria whenever possible. Even the highly confidential norm-reference tests have specifications and format samples available. These should become even more available as test developers pay attention to the Openness Standard of the *Assessment Standards* (National Council of Teachers of Mathematics, 1995): Assessment should be an open process. This standard speaks to the need for teachers and students receiving timely information of what and how information will be gathered, and that these assessments be consistent with the learning goals of the classroom.

In the mid-eighties, the Connecticut State Legislature mandated mastery testing in mathematics and language arts/writing at the beginning of grades 4, 6, and 8. These tests are developed to assess mastery of content through the end of grades 3, 5, and 7. To comply with the Openness Standard, the Connecticut State Department of Education (CSDE) developed sample item booklets and mastery test handbooks for each discipline. The CSDE has also developed comparable voluntary tests for grades 3, 5, and 7 that address the objectives complementary to those on the mandated test.

In the mathematics portion of the test, every attempt has been made to maintain assessment of developmentally appropriate content. Test items consist of multiple choice, grid-in, short answer open-response, and longer answer open-response items. The multiple choice items are gridded into an answer booklet and machine "scorable." The grid-in items tend toward "open-endedness" because the student has to calculate the correct answer and grid in the response. These items are also machine scorable. The short open-response items are scored by hand with a 2-point rubric: the answer is deemed correct (1 point) or incorrect (0 points). The longer open-response items are scored by hand with a 3-point rubric: demonstrates no or minimal understanding (0 points), demonstrates partial understanding (1 point), and demonstrates full and complete understanding (2 points).

Although it is imperative that the teacher provide the pro-am for multiple choice and grid in items within classroom instruction, the real practice for students comes in the form of the open-response items. The teacher should provide this practice to students throughout the year, not just the week before the test, as is the case with the golf tournament. This practice can have many forms. Let's look at some teaching and assessment strategies that naturally lead to success on a high stakes, externally imposed test.

Successful Strategies

On the Connecticut tests, open-response test items may ask students to show their complete work and to justify, in writing, their answer by stating a rule or explaining their thinking. Examples of these kinds of items include: finishing a numerical or geometric pattern, writing a story problem that can be solved by a given open number sentence, identifying missing or extraneous information in order to solve a problem, completing a graph from given data, providing an estimated result of a numerical operation (including whole numbers, decimal, and fractions), estimating and measuring lines, perimeter and area, drawing or describing a given geometric figure, and drawing a reflection or line of symmetry.

Good teaching strategies, which can be classified as the pro-am to a test, dictate that teachers ask "Why?" "How did you decide that?" "How else could you describe it?" after all responses, not just incorrect ones. It is through the justification of correct answers that students become aware of the need to be able to state their thinking or decision making.

For example, students - especially in the primary grades - are often only engaged in reading, completing, or creating a pattern. These patterns are usually geometric, seldom numerical. Only sometimes are they asked to describe a pattern. In other words, students need more exposure to a variety of patterns, including number patterns, and they need to be asked to state the generalization that is used to determine a pattern. This kind of preparation is truly a pro-am activity for testing at any level.

Writing word (or story) problems is not always easy; it takes a creative mind to develop problems that capture students' imaginations. Textbooks are not known for their creative story problems, but they can serve as a starting point for teachers. These sample problems can be modified by changing names or settings that are more relevant to students. They can also be used to delete information or add additional, irrelevant information in order to provide practice for students. The teacher can ask students "What else do you need to know to solve this problem?" or "Is there extra information that you do not need to solve the problem?" In time, these problem situations can move from oral to written form to prepare for an assessment. The example in Figure 1 has possible score points of 0 and 1, as shown in Figure 2.

Susie and Joey bought a bag of marbles to share. Susie chose the 12 blue marbles. Joey kept the rest.

Write what else you need to know to determine the number of marbles Joey kept?

Figure 1.   Sample Problem: Determining Missing Information

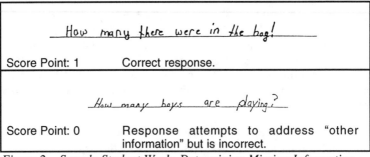

How many there were in the bag!	
Score Point: 1	Correct response.
How many boys are playing?	
Score Point: 0	Response attempts to address "other information" but is incorrect.

Figure 2.   Sample Student Work: Determining Missing Information

The pro-am of graphing must begin with the construction of real graphs that are then translated into representational graphs. Only after basic graphing skills, such as labeling and constructing the axes, are mastered should youngsters be expected to display numerical data in appropriate picto-, bar, or line graphs in an assessment task. The example in Figure 3 has possible score points of 0, 1 and 2 (Figure 4).

Mrs. Dolan collected the following information from her fourth-grade students about their favorite toothpaste.

Complete the BAR graph using the following information:

Favorite Toothpaste	Number of Students
Brighto	9
Clean White	6
Shine	5
Sparkley	7

Figure 3. Sample Problem: Graphing

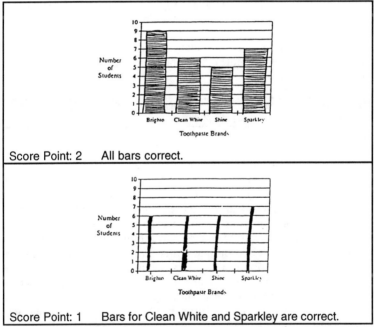

Score Point: 2   All bars correct.

Score Point: 1   Bars for Clean White and Sparkley are correct.

Figure 4. Sample Student Work: Graphing (continued on next page)

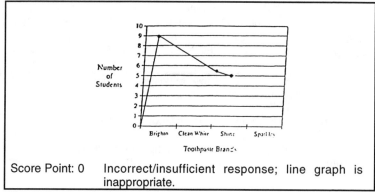

Score Point: 0   Incorrect/insufficient response; line graph is inappropriate.

Figure 4.   Sample Student Work: Graphing (concluded)

To assess the concept of the basic mathematical operations, students could be asked to write a story (word) problem which could be solved using a given number sentence (Figure 5). The first answer (Figure 6) is the classic example of a student demonstrating understanding of the addition sentence. The second example, although not as concise, demonstrates the same complete understanding of the addition sentence.

> *Write a story problem that can be solved using the number sentence*
> $9 + 4 = ?$

Figure 5.   Sample Story Problem

Simple identification of geometric shapes is not enough to serve as the pro-am for developing a real understanding of geometry. Students must routinely engage in describing and comparing shapes and identifying similarities and differences. Although young children are able to recite or mimic the teacher when naming the various pattern blocks, this must be followed by describing the properties of the shapes, albeit in age appropriate terms, to prepare them to make the necessary discrimination among shapes and internalize the accompanying properties (Figure 7). Continued practice across the grades should improve students' understanding of geometric properties which will lead to enhanced performance on geometry related questions on any test.

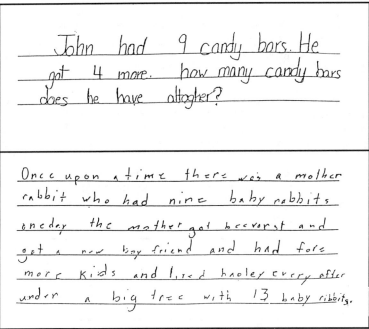

Figure 6. Sample Student Work: Story Problem

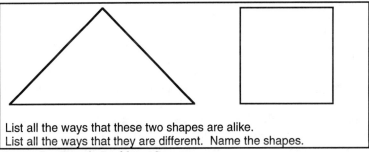

List all the ways that these two shapes are alike.
List all the ways that they are different. Name the shapes.

Figure 7. Sample Problem: Geometry

In a developmental sequence, measurement must begin with non-standard units to develop the conceptual understanding of standard measure, and to develop a purpose for using standard measuring devices such as rulers or meter sticks. Students should be able to compare measures made with Unifix cubes to measures made with pencils as more shorter units vs. fewer longer units. They may even be able to make preliminary conversions such as, "If about eight Unifix cubes equal the length of one pencil, then three pencils will equal about 24 Unifix cubes." The teacher can ask: "How are these measures alike? How are they different?" "How can we make sure that everyone's measure of a certain length means the same thing?"

The pro-am in the classroom could have students working with "broken" rulers (Figure 8) to fully develop the understanding of standard measures shown on the measuring tool. Avoid telling students "to put the edge of your object flush with the edge of the ruler." Realistically, many rulers have worn edges or a small additional space to avoid the worn edge and students must be proficient in using them accurately. This misconception (or misteaching?) became evident as students mismeasured on the Connecticut Mastery Test using a punch-out ruler provided on the test.

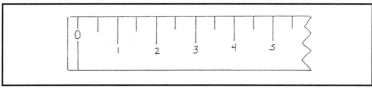

Figure 8.   "Broken" Ruler

To develop the concept of area, a teacher could ask: "How many color tiles will cover this space?" "How did you figure this out?" Since color tiles are one-inch squares, this is a form of the pro-am for developing understanding that area is reported in square units. Teachers should avoid introducing the formula, found in many 4th grade textbooks, until students have mastered these area concepts.

Measurement needs to expand to incorporate two and three dimensions. Having students "walk" around objects to determine the distance around will develop a better understanding of perimeter. Covering surfaces with tiles or blocks, as discussed above, will promote understanding of area, including area of irregular shapes. Filling 3-dimensional shapes with blocks or other items will enhance understanding of space and volume. For example, filling rectangular boxes with cubes will promote understanding of capacity and volume. These kinds of activities will build a sound platform for later use of formulas that students will, undoubtedly, face on tests.

The Standards encourage students to estimate answers, especially to check the reasonableness of their calculations. One 1986 NAEP multiple choice assessment item asks 13-year-olds to estimate the answer to $\frac{7}{8} + \frac{12}{13} =$. Only 23% chose the appropriate answer of 2. The other responses were distributed among the answer choices of 1, 19, and 21, with over 50% choosing 19 and 21. The concept of fractions is especially difficult. Teachers need to provide ample opportunities for students to engage in many estimates of the size of fractions before computing with fractions. The question, "How do you know that your answer makes sense?" should be asked often.

In this case, the pro-am might look something like Figure 9. Make "fraction cards" such as $\frac{1}{5}$, $\frac{7}{8}$, $\frac{7}{15}$, $\frac{43}{100}$, and so on, and have students sort these cards into the appropriate category. With every placement of a card, the student needs to verbalize, perhaps to a partner, the reason for placing the card in that category. Students very quickly in tune into the approximate size of various fractions, including placing cards on the lines between given choices, and, thus, are better able to make critical estimates and predict a reasonable answer.

close to 0	close to $\frac{1}{2}$	close to 1

Figure 9. Fraction Sorting Grid

Types of Assessment

What types of assessment strategies provide the most, or perhaps the best, feedback to the teacher? Among the most valuable classroom assessment strategies are teacher observation, student interviews, written assessments, math journal writing, and portfolio reviews. All of these methods provide immediate and concrete feedback to the teacher and, in turn, to the student. The teacher is encouraged to keep accurate, dated, records based on observations and interviews. These serve well to inform parents and even pupil placement teams (or child study teams) of students identified as being "in need" . By constantly asking "Why did you do ...?" "How did you do ...?" "What made you decide ...?" "Can you do it another way?" helps the teacher to answer the purposes of assessment.

- *Purpose 1*: Are students progressing? Are they becoming more informed? Are they able to transfer knowledge from one setting to another?
- *Purpose 2*: Am I, the teacher, providing the expectations, the setting, the motivational activities that help students learn and move toward being life-long learners? What, or how, must I change my instruction to improve student learning?

Conclusion

Assessment comes in many shapes and sizes. In the classroom, it can come in the form of oral questions, observing students at work or during performances, reviewing written work, responding to math journal entries, or reviewing portfolio entries made by the student. All of these are the "walking the course, testing the lie of the ball, perfecting the swing" of the pro-am.

No professional golfer would presume to enter the tournament without practice. No student, regardless of age or grade level, should have to enter the assessment tournament, especially externally imposed tests, without the same dedicated practice. Teachers, like the caddies on the tour, are the guides for our emerging "pros."

Reference

National Council of Teachers of Mathematics. (1995). *Assessment standards for school mathematics*. Reston, VA: Author.

Thoughts on Assessment in the Mathematics Classroom

Norman L. Webb
University of Wisconsin

Classroom assessment is similar to large-scale assessment in many ways. All assessment is gathering information about students to make decisions. Classroom assessment is no different. What information a teacher collects about students in a classroom will vary from what is produced by a district-mandated test, but the general purposes for both are essentially the same:

1. Sort students (e.g., assign a grade as compared to select for college admission)
2. Certify students (e.g., determine understanding as compared to judging the attainment of knowledge for graduation)
3. Diagnose students' needs (e.g., identify precise learning needs as compared to isolate deficiencies in a curriculum)
4. Evaluate instruction (e.g., judge the value of instruction for a class as compared to determining the effectiveness of a district's mathematics program)

The five major components of any assessment will vary some in nature, but still are readily identifiable in the classroom (Webb & Coxford, 1993).

1. A question, some stimulus, or an opportunity to observe
2. A response by a student or observation of performance
3. A score or consolidation of information
4. An interpretation of the score or giving meaning to the score and response
5. A report or filing of results

The time frame between seeking information and processing information is greatly reduced in the classroom in contrast to large-scale assessment. However, a teacher's asking a student a question and then filing in memory an analysis of the student's response has the same features as any other form of assessment.

Every form of assessment has to meet general criteria of quality. Reliability and validity are the two most commonly used criteria of quality. With the advent of systemic and standards-based reform, alignment is a third criteria that is becoming increasingly important for assessment (Webb, 1997). An assessment is reliable if responses are consistent. In the classroom, attending to issues of reliability is related to assuring the convergence of information from a multiple sources of information in drawing inferences about students' learning. Validity relates to assuring inferences made about a students' understanding of a concept or procedure are based on the full extent of the students' actual knowledge of the concepts or procedures being measured. Alignment of assessments with expectations requires that the assessment be directed toward students attaining important knowledge as specified by the district or state standards. Assessment aligned with expectations also will work toward having students be prepared for learning the demanding mathematics in future grades.

Even though classroom assessment has many common features with any assessment, it is different from other forms of assessment in important ways. The time frame between receiving information from a student and reporting feedback is greatly reduced in classroom assessment. Sometimes the lapse of time is nearly instantaneous between a student's response to a probing question and a teacher's insightful guidance. Along with these occasions of "reaction" assessment, over the course of the school year, teachers have the benefit of observing students for an extended period of time and getting to know in some depth the mathematical growth of each student. This differs greatly from large-scale assessment that is confined to one to three hours of student labor.

Assessment in the classroom can be done with a greater variety of techniques than can be employed under the rigorous conditions necessary for large-scale assessments. Observations, interviews, journals, portfolios, two-stage tests, take-home tests, student-constructed tests, computer-generated monitoring, and projects are a few of the techniques that classroom teachers have at their disposal to gather information from students about what mathematics they know and can do. The vast number of techniques available to classroom teachers give them the necessary tools for deeply understanding how students are making progress toward important learning goals and what interventions are needed to guide students.

Classroom assessment, compared to large-scale assessment, is not as dependent on assuring or establishing reliability or consistency of information collected. Because of the many sources of information teachers have available to them, they are more able to verify if the developing picture of a student's mathematical understanding is accurate. For example, a student's erratic responses to homework, classroom work, and group discussion on similarity all give clues to how a student is perceiving the relationship between similar figures. These independent observations, along with some well directed questions, increase the confidence of the teacher that the student has not developed a full understanding of similarity. Increasing the level of confidence about a teacher's understanding of a students' learning is more important in a classroom setting than only assuring internal consistency for any individual assessment instrument or procedure.

Effective Assessment for the Mathematics Classroom

Assessment in the mathematics classroom is one of many skills teachers have to develop and refine in gaining the content pedagogical knowledge required for successful teaching. Yet, teacher education institutions and professional development experiences give very little attention to teachers developing their understanding of what is effective assessment and how to use the full range of techniques that are available to teachers. There are some indications that while teachers value learning about assessment practices in addition to tests, preservice educational measurement courses attend primarily to statistics, preparation of exams, and the administration and scoring of exams (Gullickson, 1986). Because of these and other conditions, tests and quizzes dominate teachers' assessment practices used in classrooms and assigning grades, particularly at the high school level (Senk, Beckman, & Thompson, 1997; Stiggins & Bridgeford, 1985).

Research supports the view that teachers consider assessment important, that they are engaged in assessment activities a large percentage of their working day, and that teachers impact student learning through their assessment and evaluation practices, although there maybe a differential effect according to the achievement level of students (Webb, 1992). Teachers use a variety of assessment techniques, but the paper-and-pencil test is the dominant practice in upper-level mathematics classes.

Little research has been done to investigate the full extent of the actual assessment practices in the mathematics classroom and what impact a variety of assessment techniques can have for improved learning. A range of questions exist to be answered by research on assessment in the mathematics classroom:

1. What is evidence that students have gained understanding of specific mathematical concepts and procedures?
2. What is evidence that students have gained mathematical reasoning, communications, and facility of problem solving as expressed in the *Standards* (National Council of Teachers of Mathematics, 1989)?
3. What is evidence that the students are making progress toward achieving mathematical power as envisioned in the *Standards*?
4. How do different assessment practices and their accumulative effect add value to instruction?
5. What is the necessary balance between a teacher gathering information about a student's understanding and intervening by giving instruction and feedback to the student?
6. How can teachers be sure, in providing instruction and feedback to students, they are acting on relevant and valid information?

Supports learning. Assessment remains a means to improved learning and not an end in itself. Effective assessment practices have to be judged ultimately on how the information gathered through a variety of techniques supports learning. Question 4 above is central. Effective classroom assessment will add value to student learning.

Uses multiple techniques. There are many reasons why classroom assessment has to rely on a multiple of techniques. The knowledge a student will demonstrate on an assessment can vary by the form of assessment (Baxter, Shavelson, Herman, Brown, & Valadez, 1993). Understanding mathematics is complex and requires making connections between ideas, facts, and procedures (Hiebert & Carpenter, 1992). Different questioning techniques and different ways for students to demonstrate their growing understanding of mathematics are needed simply because of the complexity of mathematics as a body of knowledge.

Conceptually based. Effective classroom assessment needs to be conceptually based and linked to a structure of knowledge as we best know it from research on learning and expressed expectations. As more districts and states strive towards standards-based systemic reform (Massell, Kirst, & Hoppe, 1997), assessment practices, including those used by classroom teachers, need to be aligned with common goals and work toward attaining a coherent system. Also, teachers engaged in assessing student learning need to have some means of situating the student's current understanding and thinking to the organization of mathematical ideas being learned, the logical sequence in growth in understanding, and the maturation of (or ultimate) understanding of the ideas. The Cognitive Guided Instruction (CGI) Project mapped the structure of knowledge to help teachers understand possible strategies students may use and what strategies represent more advanced understanding (Fennema, Franke, Carpenter, & Carey, 1993). Teachers in Japan spend a considerable amount of time developing lesson plans to identify all the possible approaches students may use to a problem that may be the focus of one class period. The identified approaches then are used by the teachers to recognize what students do in class and to understand better students' knowledge of mathematics (Nagasaki & Becker, 1993). Some districts in the United States have developed performance indicators to help teachers better situate what mathematics they see students doing and how this mathematics relates to expected progress in learning (Beyer, 1993). Dynamic assessment is another way of thinking about conceptual frameworks for interpreting students' understanding (Newman, Griffin, & Cole, 1989). Based on Vygotsky's zone of proximal development, dynamic assessment includes giving a child a task and observing how much and what kind of help the child needs in order to complete the task successfully.

Tracks and reports students' progress. Critical to classroom assessment is monitoring the growth in students' understanding and what progress students are making. Effective classroom assessment will provide students with relevant information about what students know, what is important for students to know, and how students are making progress towards what is required for their deep understanding of mathematics. Grades based on point systems representing completion of work, attendance, and compliance fall short inadequately representing status reports of students' learning. Some high school teachers have struggled with redefining what a grade of "A" or grade of "B" means in relationship to evidence of growth in mathematical understanding (National Council of Teachers of Mathematics, 1995, pp. 57-58). In redefining letter grades to represent levels of mathematical understanding, an "A" could indicate complete understanding of concepts with evidence of extending this understanding to a range of situations. What is important for effective assessment practices is for the feedback and information that students receive to represent what mathematical progress they are making.

Valuable Skills for Classroom Assessment

As in teaching, classroom assessment requires a number of skills. Some of these skills are briefly described here. This list includes some of the important skills, but should not be construed as all of the important skills needed.

Knowledge of students' thinking. Assessment is a highly inferential practice. Students provide bits and pieces of what they know and can do in their explanations to questions, written responses to problems, and actions. For a teacher to infer from observed information what students know is aided by her having both a coherent understanding of the mathematical domains and of student's thinking within these domains (Fennema, Carpenter, & Franke, 1992). There is mounting research that indicate teachers who grasp how basic mathematical ideas develop in children are better prepared to make sense of students' work, assess their students' knowledge more often and use a greater range of assessment procedures (Carey, Fennema, Carpenter, & Franke, 1995; Fennema, Franke, Carpenter, & Carey; 1993). Students' thinking as a focus of assessment and instruction helps to develop a classroom environment with a culture of sense-making (Schoenfeld, 1989).

Effective questioning. Asking effective questions that elicit informative responses from students is critical to classroom assessment. What is an effective question will depend on the purpose for asking the question. Wilson (1996) identified six types of questions. Fact questions ask for a single, direct, and factual answer. Funnel questions encourage the student to "fill in the blank." How questions ask about a student's derivation of a solution or answer. Why questions seek the mathematical reasons behind a response or idea. Probing questions are a series of inquiries each one seeking more in depth information. Higher-order-thinking questions require an analytical or abstract response. In a case study of a grade 8 algebra teacher's assessment practice conducted by Wilson, the teacher asked different types of questions under different situations, for example, while monitoring observations of students and while engaged in direct instruction. Questioning is an essential assessment tool and art that requires gauging the type of question for the purpose at hand.

Constructing tasks. There is a preponderance of mathematical tasks used in all levels of assessment including fixed-choice, open-response, open-ended, authentic, contextual, and extended to name a few. However, developing a task or a sequence of tasks that can be used to make valid inferences of a students' knowledge of a specific mathematical concept, procedure or idea is not always easy. Developing a task requires identifying the specific assessment objectives, revealing the extent to which a student possesses knowledge of the idea being assessed, and disclosing how students have integrated their knowledge of this idea to other ideas or contexts (Webb & Briars, 1990). Developing tasks for large-scale assessments frequently requires a number of cycles of piloting and revising the task. This is particularly true for open-response questions where students are asked to generate and write the answer to a problem with one correct answer and open-ended questions where there is more than one way of stating the answer. Tasks for classroom assessment can be more robust and do not require as much refinement because it is easier to correct a poorly worded question or to gather additional information.

From a review of research, Van den Heuvel-Panhuizen (1996) identified characteristics of good assessment problems -- problems should be balanced, meaningful and worthwhile, involve more than one answer and higher order thinking, elicit the knowledge to be assessed, and reveal something of the process. She noted, as do others, that a variety of mathematical tasks are appropriate and no one form is the best (Clarke, 1993; Lamon & Lesh, 1992).

The amount of scaffolding in a task -- the number of leading questions, the amount of information given, and the format -- is an on-going issue for designing performance assessment and open-ended tasks. Sometimes breaking a complex question into parts increases the accessibility to a larger number of students and the number of students who will be able to respond to the question. With other tasks, asking only one question will produce evidence of a deeper understanding of mathematics from only a few students while inhibiting many other students from demonstrating fully what they know and can do about what is the measurement intent of the task. For example, asking students to find the nth term of a pattern can produce evidence of a significant understanding of mathematics by some students that would not be generated if the students were asked first to find the 4th term, then the 10th term, and finally the nth term.

Another important consideration for making valid inferences from students' work on classroom assessment tasks, as for any assessment tasks, is to limit students producing the right answer for the wrong reasons. The *Standards* (NCTM , 1989, p. 194) illustrate this using a task asking students to give the perimeter of a hexagon, given measures of each side. Some students may compute the correct perimeter simply because adding the six numbers is the most reasonable operation to perform with the information given and size of numbers. Successfully answering this question does not imply the student has an understanding of perimeter or could produce a figure with a given perimeter. Getting the right answer for the wrong reason can be a large problem when using multiple choice tasks by students selecting the correct choice without really understanding the underlying mathematics being tested by the task (Gay & Thomas, 1993).

Managing information. Working with up to 150 or more students a day can be overwhelming. Teachers have an administrative nightmare to keep even simple records of each student's test scores, much less recording the growth in knowledge each student is making on specific concepts. Successful classroom assessment requires some form of a management system to organize, store, and retrieve information on students' learning. There are the traditional forms of marking in grade books, recording grades, and writing down impressions. Teachers are more frequently using methods including computers for recording comments and information, a performance indicator sheet (Beyer, 1993; Meisels, Dichtelmiller, Dorfman, Jablon, & Marsden, 1993), note cards, and a "roving" hand-held tape record passed from student to student for each to record their thinking or specific problem areas (Wilson, 1996). Many recording and reporting techniques and systems exist. Those techniques focused on the mathematical ideas and content to be learned rather than only tracking participation or relative grades are more useful for assessing students' progress. Critical to developing management systems, and assessment in general, is distinguishing monitoring evidence of students' progress through a curriculum from evidence of student's developing mathematical knowledge (Shafer, 1996).

Sampling students' work. How much evidence of student learning is enough? When do I know the student is fluent in a mathematical concept? Only on narrow, precisely specified mathematical concepts and procedures (e.g., two plus two is four) can we be reasonably sure that a student knows the mathematics and will be able to use the mathematics under most conditions. Because very little mathematics of importance can be so specified, the goal of determining through assessment what mathematics a student knows and can do requires structuring situations for students to reveal their knowledge, observing, and asking questions. But even with converging information from using multiple forms of assessment, judgments still will be based only on partial evidence or a sampling of evidence. Determining the adequacy of a sample of student's work to make the needed judgments depends on factors such as the consequences of the judgments (e.g., for a final grade), the mathematical domain being assessed, what other information is available, and the cognitive and mathematical demands of the knowledge being assessed. Assessing students' knowledge of the concept of a rational number requires selective sampling to assure the student can describe, represent, and apply the full range of different forms -- common fraction, repeating decimal, percent, ratio, etc. (NCTM, 1989, p. 226). As with other assessment issues, determining what sample of student's work is necessary to reliably judge his or her knowledge requires good understanding of the mathematical ideas, the essential concepts and procedures, the relation of these essential elements with each other and to other parts of mathematics, and stages in students' growth of the central idea. It is possible to over sample in assessment by continuing to ask students to complete tasks that we already know they can do. Good sampling in assessment then is an issue of balance between gathering just the right amount of information for the purposes at hand without unneeded redundancy.

Constructing rubrics and scoring schemes. The intimacy of classroom assessment frequently allows teachers to give student immediate feedback on the quality of their work. When a record needs to be kept, when information needs to be aggregated from a number of assessment events, or when information may need to be verified at a later time, some means to summarize students' work is required. The increased use of alternative assessment tasks -- including open-ended tasks, performance assessment tasks, and portfolios -- requires scoring schemes that go beyond just assigning partial credit and adding points. Rubrics, rules for assigning a value to student work, have been successfully used in scoring students' work on mathematics tasks (Stenmark, 1991; New Standards, 1997). Analytic and holistic schemes are two common forms of rubrics. Analytic rubrics provide rules for assigning point values for specific features of the student's work. Holistic rubrics describe means of assigning a single value taking into consideration all of the student's work. Rubrics and other scoring schemes are as applicable in classroom assessment as in large-scale assessment. Important to defining rubrics in either assessment situation is setting clear cutpoints needed to classify levels of performance on the mathematical activity. Even with clear cutpoints, tradeoffs have to made in classifying performances with features of two adjacent levels.

Collaboration with others. Raising expectations in mathematics, changing curriculum emphases on emerging topics such as probability and statistics, and introducing new curricula (Fendel, Resek, Alper, & Fraser, 1997; Ohanian, 1997) place increasing demands on teachers and their assessments. Judging student's knowledge of mathematics is complex. Discussing student work with others is important for analyzing the meaning that can be given to what a student has done and to understand how a student's knowledge of mathematics is progressing. Reflection is needed to make valid inferences of student's work on complex tasks. Two people analyzing the same mathematical activity may see different important mathematical understandings demonstrated by students (Shafer, 1996). As with other instructional endeavors, collaboration in developing tasks, analyzing student work, and developing scoring schemes has value.

Classroom Climate Conducive to Good Assessment Practices

Good assessment is closely linked to good teaching. As effort is made to structure classroom experiences and its arrangement to create an effective learning environment, similar effort is necessary to build, or embed, effective assessment into teaching. Assessment should not just be the jurisdiction of the teacher, but also should engage students in making judgments about their own work and the work of their peers. Creating a classroom environment conducive to student self-assessment will require students becoming more familiar with what is quality work. They will need to become more sensitive to evidence that they understand concepts and procedures and are able to apply reasoning and do problem solving. Some teachers have had students engage in developing their own rubrics to facilitate this kind of reflection by students.

A safe environment is required for a assessment to be effectively used in a classroom. Clearly students and teachers need to feel physically safe, but they also need to feel comfortable to reveal their thinking about mathematics. Conditions need to be such that students are encouraged to talk about mathematics, open to the reflection by others on their thinking about mathematics, and actively seek feedback from others.

Above all else, assessment needs to be kept in perspective. There should be a balance between the effort required to gather and record information on students and the effort to construct effective instruction for students. An over emphasis on assessment can be a detraction to student learning. What is the appropriate balance will depend on many factors. One indication that the balance has gone too much over to assessment is if more time is spent marking student work without sufficient time to reflect on the results and plan instructional activities.

There is no question about the importance of effective classroom assessment. But we still have much to learn about how we can become better assessors of student learning. What is clear, at least from written tests and quizzes commonly used in classrooms, is that we in this country tend to assess isolated skills at levels far below expressed expectations. There is some indication that progress has been made in improving assessment practices. Assessment is becoming more aligned with high expectations. However, more remains to be done. Research on assessment practices has a great potential to serve an important role for increasing our understanding and skills in using information about students effectively to enhance their learning of mathematics.

References

Baxter, G. P., Shavelson, R. J., Herman, S. J., Brown, K. A., & Valadez, J. R. (1993). Mathematics performance assessment: Technical quality and diverse student impact. *Journal for Research in Mathematics Education, 24,* 190-216.

Beyer, A. (1993). Assessing students' performance using observations, reflections, and other methods. In N. L. Webb & A. F. Coxford (Eds.), *Assessment in the mathematics classroom: 1993 yearbook* (pp. 111-120). Reston, VA: National Council of Teachers of Mathematics.

Carey, D. A., Fennema, E., Carpenter, T. P., & Franke, M. L. (1995). Equity and mathematics education. In W. G. Secada, E. Fennema, & L. B. Adajian (Eds.), *New directions for equity in mathematics education* (pp. 93-125). New York, NY: Cambridge University Press.

Clarke, D. J. (1993, March). *Open-ended tasks and assessments: The nettle or the rose.* Paper presented at the National Council of

Teachers of Mathematics, Research Presession to the 71st Annual Meeting, Seattle, WA.

Fendel, D., Resek, D., Alper, L., & Fraser, S. (1997). *Interactive mathematics program: Integrated high school mathematics: Year 1.* Berkeley, CA: Key Curriculum Press.

Fennema, E., Carpenter, T. P., & Franke, M. L. (1992). Cognitively guided instruction. *National Center for Research in Mathematical Sciences Education. research review: The teaching and learning of mathematics* [newsletter of Wisconsin Center for Education Research, University of Wisconsin - Madison], *1*(2), 5-9,12.

Fennema, E., Franke, M L., Carpenter, T. P., & Carey, D. A. (1993). Using children's mathematical knowledge in instruction. *American Educational Research Journal, 30,* 555-583.

Hiebert, J., & Carpenter, T. P. (1992). Learning and teaching with understanding. In D. A. Grouws (Ed.), *Handbook of research on mathematics teaching and learning* (pp. 65-97). New York, NY: Macmillan.

Gay, S., & Thomas, M. (1993). Just because they got it right, does it mean they know it? In N. L. Webb & A. F. Coxford (Eds.), *Assessment in the mathematics classroom: 1993 yearbook* (pp. 130-134). Reston, VA: National Council of Teachers of Mathematics.

Gullickson, A. R. (1986). Teacher education and teacher-perceived needs in educational measurement and evaluation. *Journal of Educational Measurement, 23,* 347-354.

Lamon, S. J., & Lesh, R. (1992). Interpreting responses to problems with several levels and types of correct answers. In R. Lesh & S. J. Lamon (Eds.), *Assessment of authentic performance in school mathematics* (pp. 319-342). Washington: American Association for the Advancement of Science.

Massell, D., Kirst, M., & Hoppe, M. (1997, March). *Persistence and change: Standards-based systemic reform in nine states* (Policy Briefs, RB-21). Philadelphia, PA: Consortium for Policy Research in Education, Graduate School of Education, University of Pennsylvania.

Meisels, S. J., Dichtelmiller, M., Dorfman, A., Jablon, J. R., & Marsden, D. B. (1993). *The work sampling system resource guide.* Ann Arbor, MI: Rebus Planning Associates.

Nagasaki, E., & Becker, J. P. (1993). Classroom assessment in Japanese mathematics education. In N. L. Webb & A. F. Coxford (Eds.), *Assessment in the mathematics classroom: 1993 yearbook* (pp. 40-53). Reston, VA: National Council of Teachers of Mathematics.

National Council of Teachers of Mathematics. (1989). *Curriculum and evaluation standards for school mathematics.* Reston, VA: Author.

National Council of Teachers of Mathematics. (1995). *Assessment standards for school mathematics.* Reston, VA: Author.

New Standards. (1997). *1996 New Standards reference examination technical summary: A report of the New Standards Technical Studies Unit.* Pittsburgh, PA: Learning Research & Development Center, University of Pittsburgh.

Newman, D., Griffin, P., & Cole, M. (1989). *The construction zone: Working for cognitive change in school.* Cambridge, England: Cambridge University Press.

Ohanian, S. (1997). Math that measures up. *American School Board Journal, 184*(6), 25-28.

Schoenfeld, A. H. (1989). Problem solving in context(s). In R. I. Charles & E. A. Silver (Eds.), *The teaching and assessing of mathematical problem solving: Volume 3: Research agenda for mathematics education* (pp. 82-92). Reston, VA: National Council of Teachers of Mathematics.

Senk, L. S., Beckman, C. E., & Thompson, D. R. (1997). Assessment and grading in high school mathematics classrooms. *Journal for Research in Mathematics Education, 28,* 187-215.

Shafer, M. C. (1996, April). *Assessing growth in a mathematical domain over time.* Paper presented at the annual meeting of the American Educational Research Association, New York City, NY.

Stenmark, J. K. (Ed.). (1991). *Mathematics assessment: Myths, models, good questions, and practical suggestions.* Reston, VA: National Council of Teachers of Mathematics.

Stiggins, R. J., & Bridgeford, N. J. (1985). The ecology of classroom assessment. *Journal of Educational Measurement, 22,* 271-286.

Van den Heuvel-Panhuizen, M. (1996). *Assessment and realistic mathematics education.* Utrecht, The Netherlands: Freudenthal Institute.

Webb, N. L. (1997). *Criteria for alignment of expectations and assessment in mathematics and science education* (Research Monograph No. 6). Madison: National Institute for Science Education, Wisconsin Center for Education Research, University of Wisconsin.

Webb, N. L. (1992). Assessment of students' knowledge of mathematics: Steps toward a theory. In D. A. Grouws (Ed.), *Handbook of research on mathematics teaching and learning* (pp. 661-683). New York, NY: Macmillan.

Webb, N. L., & Briars, D. (1990). Assessment in mathematics classroom. In N. L. Webb & A. F. Coxford (Eds.), *Assessment in the mathematics classroom: 1993 yearbook* (pp. 108-117). Reston, VA: National Council of Teachers of Mathematics.

Webb, N. L., & Coxford, A. F. (Eds.). (1993). *Assessment in the mathematics classroom: 1993 yearbook.* Reston, VA: National Council of Teachers of Mathematics.

Wilson, L. D. (1996, April). *Documenting observations of students in mathematics: A case study.* Paper presented at the annual meeting of the American Educational Research Association, New York City, NY.

Section 3

Working Papers

The Role of Teachers' Knowledge in Assessment

Carne Barnett
WestEd

Designing productive assessment questions and tasks and relating them to instruction requires considerable content and pedagogical content knowledge in mathematics (Ball, 1988; Featherstone, Smith, Beasley, Corbin, & Shank, 1995; Fennema, Franke, Carpenter, & Carey, 1993; Russell, Shifter, Bastable, Yaffee, Lester, & Cohen, 1995; Wilcox, Lanier, Schram, & Lappan, 1992). When this knowledge is underdeveloped, we are less able to choose tasks or ask questions that reveal the understandings, partial understandings, and misunderstandings of students. Likewise, we are less able to use the information we receive from tasks and questions to inform instruction in productive ways. The knowledge base that is necessary includes such things as a complex understanding of the various ways concepts can be represented and misrepresented, the oral and written language and symbols, associated problems and applications, and connections to other mathematical ideas.

In order for assessment to have a valuable impact on instruction, we must have a clear picture of what needs to be understood or learned and be able to detect students misconceptions or partial understandings. For example, if we believe that understanding a fraction such as $\frac{5}{8}$ can be demonstrated by shading five out of eight parts on a pre-partitioned circle, we will be convinced that students "understand fractions" if they can accomplish this task. Even having students draw and partition the circle themselves would provide a slightly better assessment, but still would not help us know if the student understood that $\frac{5}{8}$ might be represented in many ways, such as on a graph or as a measure on a ruler. Nor would we know whether the student has a common misconception such as thinking that $\frac{5}{8}$ is greater than $\frac{3}{4}$ because the numerator and denominator for $\frac{5}{8}$ are greater than for $\frac{3}{4}$. Finally, we would not know if students could apply the concept of $\frac{5}{8}$, perhaps to a complex data analysis problem or a scientific situation.

Although new assessment and curriculum materials are designed to broaden our understandings of what students can do and understand, this may not be enough. Considerable background knowledge, integrating both content and pedagogy, is necessary to support student learning and assessment that goes beyond simple procedures or routine problems.

It seems important to ensure that professional development on assessment focuses on both the processes to be used to assess and the content to be assessed. It is not enough to use a well crafted performance assessment, we must also understand when to use a particular assessment to measure a particular targeted outcome and what to do with the information that we acquire. Again, all of this requires deep content and pedagogical content knowledge. We need to ask ourselves: Are our assessment workshops mistakenly assuming that this knowledge is in place? If not, how might we configure professional development so that assessment issues become a bridge to better understand mathematics content?

Many teachers have been involved in analyzing student work as part of their professional development experiences in assessment. However, the depth of discussion about student work often relies on the depth of mathematics knowledge teachers bring to the discussion. I would like to propose that one promising approach to deepening the analysis of student work and strengthening instruction is through the study of cases of practice. I will speak from my experience as director of the *Mathematics Case Methods Project.*

This professional development project involves elementary and middle school teachers in discussing "cases" of practice similar to case discussions in other professions such business and medicine. A central aim of the project, supported by research (Barnett, 1991; Barnett & Friedman, 1997; Gordon & Heller, 1995), is to equip teachers with a stronger and more complex knowledge base so that they can continually assess how their teaching practices impact students' thinking and learning.

Case discussions have many similarities to professional development experiences that focus on analyzing student work. While analyzing student work is a priority in each case discussion, emphasis is also given to (a) developing the teachers' own understanding of the mathematics and (b) examining how the task, the materials, and dialogue portrayed the case can be a positive or negative influence on the student's responses. Together, case discussion participants learn how to find clues about the trouble spots and consider how to help students work through what they don't understand. A higher standard for analyzing student work and instructional planning develops as teachers gain deeper understandings of the mathematics and the impact of instruction on learning.

I am concerned that new assessment practices may fail to improve learning if teachers are unprepared to use their outcomes productively. Again, the blame will be laid on teachers. However, the current interest and support being provided for assessment is an important opportunity. Studying student work through cases may be a good starting place for teachers who are not ready to put their own students' work out for scrutiny by other teachers. It also has the advantages of relating the analysis of student work to teaching practices and enhancing teachers' own content knowledge. In all, they will be better prepared to use assessment to inform their practice.

References

Ball, D. (1988, September). *The subject matter preparation of prospective mathematics teachers: Challenging the myths* (Research Report 88-3). East Lansing, MI: Michigan State University, National Center for Research on Teacher Education.

Barnett, C. (1991). Building a case-based curriculum to enhance the pedagogical content knowledge of mathematics teachers. *Journal of Teacher Education, 42*, 263–272.

Barnett, C., & Friedman, S. (1997). Mathematics case discussions: Nothing is sacred. In E. Fennema & B. Scott-Nelson (Eds.), *Mathematics teachers in transition* (pp. 381-399). Hillsdale, NJ: Lawrence Erlbaum Associates.

Featherstone, H., Smith, S. P., Beasley, K., Corbin, D., & Shank, C. (1995, January). *Expanding the equation: Learning mathematics through teaching in new ways* (Research Report 95-1). East Lansing, MI: Michigan State University, National Center for Research on Teacher Education.

Fennema, E., Franke, M., Carpenter, T. & Carey, D. (1993). Using children's mathematical knowledge in instruction. *American Educational Research Journal, 30,* 555-583.

Gordon, A., & Heller, J. (1995, April). *Traversing the web: Pedagogical reasoning among new and continuing case methods participants.* Paper presented at the annual meeting of the American Educational Research Association, San Francisco, CA.

Assessing Procedural and Conceptual Knowledge in the Mathematics Classroom

Sarah B. Berenson, Draga Vidakovic, and Glenda Carter
North Carolina State University

Several years ago, a group of fifth-grade teachers were discussing how well their students understood the operations of multiplication and division. They were confident that nearly all of their students had mastered this subject matter. A group of university researchers asked these teachers to assess their students' understanding of multiplication and division in another way. Teachers returned to their classes and asked their students to write several word or "story" problems in which multiplication and division were used. Much to the teachers' surprise, only half of the students could write an appropriate word or "story" problem for multiplication and division (Figure 1). The teachers revised their opinions of their students' understanding. They agreed that their students understood the computational processes of multiplying and dividing. In other words, if students were given a two-digit by two-digit multiplication example, the students could use the multiplication algorithm to obtain the correct answer. However, these teachers concluded that their students had weak understanding of the concepts related to multiplication and division procedures.

Write a story problem that uses multiplication and solve.

Student Response:
Sarah had 30 jacks and she lost 3 of them and wanted the same amount she had before.

30 x 3 = 90

---

*Figure 1.   Assessing a Student's Understanding of Multiplication
(Norwood & Carter, 1994, p. 147)*

This illustration of students' mathematical ideas points to the importance of student assessment. It raises several issues to be addressed. First, teachers will want to know what kinds of student information are important in the assessment process. Second, teachers and others will want to know what constitutes quality assessment in the classroom. A final consideration of this paper is to discuss whether or not different kinds of evidence are dependent on grade, age, and other student variables.

Student Information

There are at least two types of information that teachers will want to assess. First, there is information about the students' subject matter knowledge of mathematics. Meaningful instruction can be achieved when teachers understand the extent of their students' mathematical ideas. Second, there is a need to know about the cultural diversity of the students. Information about the students' socio-cultural backgrounds will prove invaluable in understanding the real-world experiences that are brought by the students to the teaching/learning process.

Students' Subject Matter Knowledge

There has been an extended dialogue in the mathematics education community about the different kinds of mathematical content knowledge. Shulman (1986) and Hiebert (1986) are in agreement that subject matter knowledge in mathematics can be classified into two types of mathematical content knowledge: (a) procedural knowledge and (b) conceptual knowledge. Skemp (1976) uses instrumental knowledge and relational knowledge as his categorizations of subject matter knowledge. Both types of knowledge are viewed as equally important. Algorithms, formulas, and definitions are types of procedural content knowledge. Conceptual content knowledge is described as knowledge that communicates the whys or the importance of the ideas related to the procedural knowledge. Borko, Eisenhart, Brown, Underhill, Jones, & Agard (1992) describe conceptual knowledge as the knowledge of what it means to do mathematics, including the nature of mathematical thinking and mathematical discourse. Hiebert (1986) discusses conceptual mathematical knowledge as the relationships of ideas. Conceptual content that emphasizes relationships establishes the connections among mathematical ideas and their meanings. Asking which ideas are related, how the ideas are related, and what happens if certain variables are changed further enhances the assessment of conceptual meaning. Instances or examples of the idea, as well as counter-examples or non-examples, can provide additional information about a student's conceptual understanding. The use of different representations including symbols, numbers, words, tables, charts, and graphs increases the opportunities of assessing conceptual understanding. Some examples of procedural and conceptual knowledge are illustrated in Figure 2.

Procedural Knowledge	Conceptual Knowledge
formulas	representations
definitions	relationships
calculations	meanings
procedures	examples/non-examples

*Figure 2.   Sample Descriptors that Differentiate Conceptual
and Procedural Knowledge*

In the past, classroom emphasis has been placed on procedural knowledge and especially computational skills in mathematics. There seemed to be a sense that if a child can compute then the child understands mathematics. The popular press and many parents continue to adhere to this belief. However the National Council of Teachers of Mathematics (1989, 1995) recognized the need to incorporate other kinds of knowledge, beyond computational skills and definitions, into school reforms. It is no longer sufficient to gather evidence of procedural knowledge in today's mathematics classroom. Assessment reforms include the necessity of collecting evidence concerning both types of knowledge: students' procedural and students' conceptual knowledge in mathematics. Figure 3 contrasts two different kinds of assessment items: the first a traditional item to assess procedural knowledge and the second an assessment item to assess conceptual knowledge.

Teachers are advised to have high expectations that all students will learn and value mathematics. With a focus on equity, the reforms are aimed at providing all students with opportunities to gain *"mathematical power."* Yet students increasingly represent diverse communities so that a variety of cultural tools and language converge within the classroom environment. The action research, given below, illustrates the importance of assessing the multiple meanings that children from diverse cultures bring to the classroom context.

Socio-cultural Knowledge
It has been suggested that teachers introduce the concept of division with the notion *fair share.* Teachers in grades 3-8 in a large rural county asked their students to give their definitions of *share* and *fair share* within a problem solving episode of sharing 36 pieces of candy

(Berenson, in press). Some children would share the candy equally among themselves and others. Some would share only one piece of candy, while keeping the rest for themselves. One student would give all the candy to others because he could "get candy whenever I want it." Another student from an impoverished family would share two pieces each while keeping the rest of the candy for "another day." One seventh grade girl reported that her mother would get the most and her younger sister would get the least and that was an example of fair share. Experienced teachers are often aware of the various meanings that children from different cultures bring to the classroom. However, even these 11 teachers were surprised with the variety of meanings their students had for these words associated with division. Shulman (1986) refers to this type of teacher knowledge about students as pedagogical content knowledge. Researchers have noted the difficulties beginning teachers have in knowing the various meanings that students have for words used in the mathematics classroom. To establish the multiple meanings of words, assessments can be used to develop the teacher's sense of children's word meanings.

Traditional Item

Student A: Which of these numbers is even?

   A. 65   B. 41   C. 19   D . 14

Alternative Assessment Item

Student B   Is 14 an even number or an odd number?   *even*
              Explain or show why.

If you take 14 cookies and split them between you and your friend, you each get 7. That makes numbers even ... you have the same amount.

$$7$$
$$O\ O\ O\ O\ O\ O\ \underline{+\ 7}\ \ O\ O\ O\ O\ O\ O$$
$$14$$

Figure 3.   *Contrasting Assessment Items for Two Types of Evidence about Student Understanding (adapted from NCTM, 1995, p. 59)*

Types of Evidence

When collecting evidence of students' understanding, there are several principles to be considered. These principles of quality assessment advise teachers to use (a) multiple measures, (b) multiple representations, and (c) multiple formats of assessment. When teachers use multiple measures they obtain evidence about students' mathematical power over time and with a variety of formats. Pictures, words, tables, graphs, and charts are among the multiple representations that are used in quality assessments, either by the teacher or the students. There is value to standardized measures such as multiple-choice, paper and pencil tests, but these sources of evidence can be greatly enhanced with the inclusion of multiple formats within the classroom. Multiple formats include journals, interviews, performances, and portfolios.

Multiple Measures

Experienced teachers recognize that students' responses to assessments can change from one day to the next. That is, a student may give a thoughtful answer to an assessment question on one day and an incorrect answer on the next day. Another student can give a reasonable answer to a question that is worded in a certain way, but an insufficient answer if the concept is assessed with different words. The variable performance of students over time with either procedural or conceptual knowledge can be frustrating for teachers. Unfortunately we cannot fill our students up at the "knowledge gas pump." The ideas of students are continually changing from one day to the next. Incomplete ideas lead to successful performances on some items and unsuccessful performances on other related items. Teachers who take multiple measures of their students' understanding gather enough evidence for teachers to make informed instructional decisions.

Multiple Representations

One of the challenges that all teachers face is convincing their students that words are used to learn mathematics. Students can be amazingly resistant to the notion that writing an explanation is a worthwhile mathematics activity. They have become accustomed to completing numerical/symbolic exercises that require numerical/symbolic answers. Taken as one representation [numbers], this type of assessment is not sufficient to assess deeper levels of understanding. Teachers who use multiple representations have more evidence of what their students know about procedures and concepts. A word problem may require a numerical answer. A set of data (numbers) may require students to construct a table or a graph. If students are given a graph to analyze, they may have to use words, either spoken or written, to describe the graphical representation. Figure 4 illustrates several different kinds of assessment items that use multiple representations. The use of multiple representations is linked to the next quality assessment principle of multiple formats.

How could you use the Base 10 blocks to teach Sue that .1>.05>.019? Use words and pictures to explain your answer.	Below is a Golden Rectangle. The ratio of the length to the width determines if a rectangle is "Golden." Construct another Golden Rectangle that is similar to but not congruent to the Golden Rectangle below.
Write a description of a family car trip using the graph below. 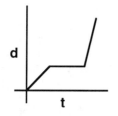	Use the CBL motion detector to demonstrate how the changes of speed are related to the distance traveled.
Find the pattern in the table below to complete the table. 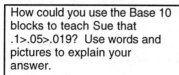	Design an investigation that explores the relationship among the angle bisectors in a triangle.
Use the graphics calculator to demonstrate different translations of the basic parabola: $y = x^2$ Explain in words, symbols, and sketches.	Construct a model of a cube using centimeter paper. Determine the surface area of your cube.

*Figure 4.   Sample Assessment Items that Illustrate the Use of Multiple
Representations for Questions and Answers*

Multiple Formats Blend Assessment with Instruction

Traditionally, classroom assessments have relied on several formats including single answer [closed response], fill-in the blanks, or multiple choice items. Students become adept at answering these types of questions after years of practice. When deviating from these familiar formats teachers have heard their students complain because they *"have to think!"* There are claims that good assessment is good instruction. A well-constructed item can provide a context for students to learn *and* think. This paper considers some alternative formats that some teachers have used successfully. They include: journals, interviews, performances, and portfolios. The questions illustrated in Figure 4 can be adjusted to fit any of the above formats.

Teachers find that to entice their students to write about mathematics that *journals* prove to be an effective assessment vehicle. Teachers in our action research groups begin with questions that assess students' feelings (e.g., Write your math autobiography since first grade.). They find that students are more than willing to share how they feel about learning mathematics as a first writing experience. After the initial entry, teachers then use journals to assess their students' understanding of mathematics. Words and pictures can be entered in students' journals at key points in the semester, so that over time, teachers can gather evidence in another format about their students' mathematical power.

Interviews provide an additional format for teachers to assess their students' thinking. Many teachers do this on an informal basis either in whole group or small group instruction. With interview formats, teachers ask probing questions of individuals or groups of students including, *What happens if ... What is the relationship ... What are some examples ... What does this mean ... How does this work ...?* An interview differs from instruction in several ways. First, the teacher does not approve or disapprove of any answers given during the interview. The purpose of the interview is to listen to the students' ideas just as a paper and pencil test records the students' thinking without the teacher's intervention. Also, each interview may differ somewhat from student to student as different probing questions are needed to assess different ideas. Second, if the interview is graded, feedback can be given to the student once the teacher has reflected on the interview. Teachers sometimes video tape or audio tape interviews to capture additional evidence that might otherwise be forgotten. Rather than interview all students, teachers interview five or six students while others are taking a paper-and-pencil test. Over the course of a month to six weeks, all students will have had an opportunity to discuss some of their mathematical ideas with their teacher.

Performance assessment formats require students to create, demonstrate, construct, build, or invent. The quality of evidence obtained from a performance assessment varies according to the type of performance that is required of the students. Teachers often set up performance centers around their room, having students rotate through different performance stations. Some performances are recorded with paper and pencil, other performances produce physical models made with tools and manipulatives. Sometimes teachers gather evidence at one of the performance stations, while the remaining students work either at their desks or at other stations. Other teachers have used video cameras to record students' performances.

The *portfolio* assessment format offers teachers and students an organized, long-term view of the students' work in mathematics. Teachers, students, parents, and other teachers can examine students' portfolios to add to their understanding about the students. There are a variety ways that teachers have their students construct portfolios. Some teachers prefer that students select their "best" work for the portfolio, while other teachers work from a list of what is to be included in the portfolio. Teachers find that portfolios are especially helpful during parent conferences in which the student makes an organized presentation to a parent, guardian, or the principal. Together with journal formats, interview formats, and performance formats; portfolios offer another opportunity for the teachers to collect quality evidence.

Grade, Age, and Other Variables

It is our view that good assessments go beyond the scope of the grade level's course of study. In metaphorical terms, mathematical power is not a linear function. This is to say, students do not all progress at the same rate nor with the same knowledge through the grades. Also the students' reading, writing, and language abilities may prove to be important variables in the assessment process. Good assessment techniques recognize this complexity and work to reveal students' ideas beyond "correct" or "incorrect" answers. These techniques provide the evidence for teachers to make instructional decisions about their students. Because mathematical knowledge is connected and continually under construction, robust items can be used across several grade levels. While the items may not change according to age or grade level, the rubrics of a particular item and the teachers' expectations will change across these dimensions.

Creating an Environment of Shared Responsibility

Quality assessments can free teachers from some of the tremendous responsibility they assume every day in the classroom. Ms. Dove's story is a good example as to how assessment can alter one's perspective. An excellent teacher of many years experience, Ms. Dove decided to experiment with different assessment techniques with her fifth-grade class. Her students had always learned and achieved in her classroom. However, those test scores and report card grades had cost Ms. Dove many sleepless nights as she pondered what to say, how to solve, and how to explain each and every math example for her students. She felt all of the responsibility for her students' learning. As she began to use alternative assessments techniques she discovered many revealing things about her students. She noted that her students were capable of interesting and complex solutions to problems. Ms. Dove's students were given the opportunity of "inventing" multiple strategies. They expressed their ideas using many different representations. Students' explanations to other students were just as understandable as Ms. Dove's. If truth be told, the students listened to one another, often with more attention than they did to Ms. Dove. These different kinds of assessments had freed Ms. Dove to teach and her students to learn.

References

Berenson, S. (in press). Language and diversity. *Focus on Learning Problems in Mathematics.*

Borko, H., Eisenhart, M., Brown, C., Underhill, R., Jones, D., & Agard, P. (1992). Learning to teach hard mathematics: Do novice teachers and their instructors give up too easily? *Journal for Research in Mathematics Education, 23,* 194-222.

Hiebert, J. (Ed.). (1986). *Conceptual and procedural knowledge: The case of mathematics.* Hillsdale, NJ: Lawrence Erlbaum.

National Council of Teachers of Mathematics. (1989). *Curriculum and evaluation standards for school mathematics.* Reston, VA: Author.

National Council of Teachers of Mathematics. (1995). *Assessment standards for school mathematics.* Reston, VA: Author.

Norwood, K., & Carter, G. (1994). Journal writing: An insight into students' understanding. *Teaching Children Mathematics. 1*(3), 146-149.

Shulman, L. (1986). Those who understand: Knowledge growth in teaching, *Educational Researcher, 15*(2), 4-14.

Skemp, R. (1976). Relational understanding and instrumental understanding. *Mathematics Teaching, 77,* 1-7.

Classroom Assessment: Translating Information about Students into Instruction

George W. Bright

The University of North Carolina at Greensboro

Although teachers are required to make evaluative judgments (e.g., grades) about their students, the primary use of classroom assessment information, as opposed to summative evaluation information, is instructional decision making. It is generally conceded that in order to student performance to improve, instruction must be designed to help students develop greater understanding about mathematics. Teachers can redesign instruction in this way if they generate detailed information about their students' thinking. At one level this understanding can develop globally. Indeed, good teachers have always been able to predict many of the common difficulties that students have in understanding mathematics. But as the population of students enrolled in "advanced" mathematics becomes more inclusive, and as the goals of mathematics instruction relate more to problem solving and reasoning, teachers need much more detailed knowledge of how individual students think in order to adapt instruction to meet the needs of those students. Teachers need to have many more tools available for probing understanding; for example, questioning, long-term projects, rich mathematics tasks, etc.

One way to monitor students' thinking is listening to students' solutions to significant mathematics problems. Teachers should expect that when students solve problems (or at least, attempt to solve problems), those solutions will make sense to the problem solver, even if the teacher does not immediately follow the logic of those solutions. But teachers can also set an expectation that students will learn how to explain their solutions so that other students and the teacher can understand the solution. Teachers should expect that students will be able to communicate their solutions coherently and that students will reflect on the quality of their solutions. Solving problems and being able to explain solutions are major components of mathematical power, a notion that has become important as part of the NCTM *Standards* documents.

In instruction on graphs, for example, several authors (e.g., Bertin, 1983; Curcio, 1987, 1989; McKnight, 1990; Wainer, 1992) have identified three levels of questions that students might be asked to help them understand and interpret information in a graph: an elementary level which focuses on extracting data from the graph, an intermediate level that involves interpolating and finding relationships in the data as shown on the graph, and an overall level that involves extrapolating from the data and interpretation of the relationships identified from the graph.

In the context of a standard bar graph, where the y-axis is labeled as frequency, of lengths of cats (Friel & Bright, 1995), these three levels are illustrated by the following questions:

1. How many cats are 30 inches long from nose to tail? How can you tell?
2. If you added up the lengths of the three shortest cats, what would the total of those lengths be? How can you tell?
3. What is the typical length of a cat from nose to tail? Explain your answer.

Although many middle grades students' responses demonstrate understanding of the information, some of these responses reveal a number of important confusions: (a) the bars may be interpreted as representing individual cats, (b) the shortest bar may be interpreted as representing the shortest cat, (c) labels on the x-axis may be interpreted as occurring in order by time, and (d) the data set may be equated with the set of labels on the x-axis. Many of these confusions were observed both before and after a unit on graphing, so they appear to be somewhat resistant to change.

Instructional decisions need to be based on assessment of students' thinking as reflected in the strategies that they use to solve problems. In addition, as students talk about mathematics concepts in a variety of contexts, they will use exemplars of key concepts. Teachers need to know whether those exemplars clearly model the essential features of mathematical ideas. The ways that students formulate arguments in support of particular solutions will also reveal much about the nature of students' reasoning. The implicit (or omitted) elements of an argument may be as revealing of students' understanding as the elements that students include in such arguments.

As the grade level increases, teachers should expect greater attention to the details of solutions and should expect more logical arguments. Too, as the grade level increases, teachers need to have more detailed frameworks of both mathematics and students' thinking in order to be able to interpret students' explanations adequately. The research base describing students' thinking is not uniform across grades or across content. Without that research base, teachers may not be able to interpret fully the sense of students' arguments.

Questioning is the most useful technique for obtaining information about students' thinking. Teachers need to think carefully about the initial questions that they ask at the beginning of a "debriefing" of solutions to a problem. But perhaps more important, teachers need to develop skill at creating probing questions that will not only reveal students' thinking but also help students to reflect on their thinking. Teachers have to create these questions in real time during interactions with students; developing skill at knowing what questions to ask does not necessarily happen quickly.

In a study of preservice elementary mathematics teachers, Vacc and Bright (in press) note the struggle of one preservice teacher during the last month of her student teaching experience in balancing "questioning" and "telling" during a lesson on perimeter. Apparently, Helen (a pseudonym) did not hear exactly what she expected in one student's response to questions, so Helen "took over" the lesson by imposing her thinking on the lesson.

> Prior to the following segment, Helen had indicated that everyone was going to use a piece of string and had asked how they would use it to determine the perimeter [of the room].

> Deanna: You could take the strings and go around the room and then take the ruler to see how long each string was, so you'd know how long the string was to count how long they are.
> Helen: Okay, to see how many inches or feet there are?... Okay, do we need to use the ceiling?
> Deanna: No.
> Helen: We can use what?

Tien: The floor.
Helen: The floor. Anywhere, really; you can use the wall. I think it would be easiest, well I don't know. It might be easier to use the wall. Whatever you want to use.... You all came up with some good ways to figure out the perimeter.... I'm going to give each two people a string ... [and] assign you a wall.... So, if you had this wall, where are you going to start.... So one partner -- I need a volunteer -- will hold it there? So Sandy is going to hold it there and I'm going to bring it around here. How many strings is the wall so far?... Okay, you let go of your end, Sandy, and bring it around the wall. How many strings is that?

Although Deanna's response indicated a clear understanding of how the room's perimeter could be measured, it appears Helen believed that she needed to demonstrate the process before the students could proceed on their own. (Vacc & Bright, in press)

Helen's comments in her long monologue at the end of this segment indicate her struggle with knowing when to question and when to tell. She seemed to want to acknowledge students thinking (e.g., "You all came up with some good ways to figure out the perimeter."), and yet she seemed to think that focusing their attention on measuring around the walls would somehow make the task easier or more understandable for the students. This particular student teacher had been mentored for almost two years by a faculty member who emphasized the importance of letting children construct their own understandings of mathematics concepts. Yet, Helen "could not resist" putting in her own two-cents worth as a means of trying to help students by "short circuiting" their own struggles with attaching meaning to the concept of perimeter.

Of course, asking questions requires teachers to listen very carefully to how students respond to those questions. Interpreting what students are intending to say, in addition to what they actually say, is often difficult to do within the rapid-paced environment of a classroom. It is often only upon reflection on a response that teachers can make complete sense of that response. Teachers need time outside of direct instruction to reflect on their questions and to conference with other teachers in developing better questioning skills.

There are other techniques that may help teachers understand students' thinking. Statistics education is a particular active area in which alternative strategies, such as projects, small group work, questioning, and "real-life" problems, are being investigated. (See Gal & Garfield, 1997, for a discussion of these.) Projects, for example, reveal ways that students' solutions develop over time as well as processes that students use to relate many different concepts that may or may not seem immediately to be relevant for solving a particular problem. Projects might be especially valuable at the middle school and high school levels, since by that time students have a wide-enough repertoire of mathematics concepts to tackle significant projects. Similarly, portfolios seem to offer some hope of seeing change over time, but there does not yet seem to be enough known about how teachers can manage portfolio use or about how teachers might interpret the information contained in portfolios. Interviews of individual students can also reveal much information about students' thinking. Even though interviews take a lot of time to conduct, they may have tremendous payoff for teachers in developing understanding of students. Video tapes of interviews or of classroom instruction may also help teachers. For example, the Annenberg/CPB Math and Science Collection (e.g., WGBH, 1996) and the PBS MathLine series have both used videotapes of classroom instruction as a vehicle for initiating discussion about how to improve teaching and understand more about students' thinking. Curriculum materials (e.g., teacher's edition of a textbook) might be restructured to provide important information to teachers about how to assess students thinking or about how students generally approach the task of solving particular problems.

One way for teachers to organize the information they obtain about students' thinking is by understanding frameworks of content and students' thinking. These frameworks generally evolve from deep thinking about the content and from research on students' thinking. Cognitively guided instruction (CGI) provides an example of such frameworks. In the process of implementing CGI (e.g., Fennema, Franke, Carpenter, & Carey, 1993), teachers are given opportunities to understand one framework of problem types and another framework of students' solutions to those problems. As teachers struggle with understanding and using these frameworks, they have to rethink their perspectives on teaching (e.g., Bright, 1996).

In geometry, the van Hiele levels of thinking is a possible framework that teachers might find useful (e.g. Fuys, Geddes, & Tischler, 1988), and for graphing, there are frameworks emerging for both the mathematics content and for students' strategies for interpreting information in graphs (Bright, Curcio, & Friel, 1996; Friel, Bright, Frierson, & Kader, 1997). In other areas of mathematics, the potential frameworks are not so easily identified. There is quite a bit known about students' thinking in areas like fractions, ratio/proportion, and functions, but there are not clear frameworks for any of these areas that tie all of these bits of information together.

Teachers in projects like CGI, Project IMPACT (Campbell & Robles, 1997), and QUASAR (Silver & Smith, 1996) provide models of ways that teachers can be supported as they gradually adjust to making instructional decisions in ways that respond to the needs of individual students. It is clear in many projects which are trying to help teachers make better instructional decisions (see Friel & Bright, 1997, for examples) that considerable time and extensive support are needed for teachers. There is no algorithm for helping teachers make this kind of substantial change; teachers and staff developers must struggle together to understand what information is important for teachers to have about students' thinking and then how to make use of that information to make effective instructional decisions.

When teachers act in a judgmental way about the quality of students solutions, students soon realize that they do not need to learn how to reflect on their solutions and how to determine which of several different solutions discussed in class might be better or more efficient. Instead teachers need to help students learn to evaluate solutions themselves to determine which ones are most reasonable and which ones are most efficient. Students need to learn to listen to each other and to ask questions that will help them to understand other students' points of view.

Similarly, teachers should help parents understand that there are different expectations for students' learning today than probably existed for the parents a generation ago. This might be communicated if teachers share students' solutions to problems so that parents come to understand the level of sophistication of the mathematics that students are expected to master. Simply telling parents is likely not to communicate the entire story.

Concluding Comments

Translating information about students into instruction is very complex. Teachers need to internalize a great deal of knowledge, including but not limited to a variety of frameworks of content and students' thinking. The process of creating this extensive knowledge base is one that never ends; teachers continue throughout their entire careers to add to their knowledge, sometimes from reading, sometimes from interacting with colleagues, but mainly through interacting with students and reflecting on what they say and do. The "good news" is that with adequate support, teachers can develop this knowledge and learn to use it in instructional decision making. When this happens, teachers are rejuvenated professionally and their students learn more.

References

Bertin, J. (1983). *Semiology of graphics* (2nd ed., W. J. Berg, Trans.). Madison, WI: University of Wisconsin Press.

Bright, G. W. (1996). Understanding children's reasoning. *Teaching Children Mathematics, 3*(1), 18-22.

Bright, G. W., Curcio, F., & Friel, S. N. (1996, April). *Building a theory of graphicacy: Where are we now?* Paper presented at the Research Presession for the annual meeting of the National Council of Teachers of Mathematics, San Diego, CA.

Campbell, P. F., & Robles, J. (1997). Project IMPACT: Increasing the mathematical power of all children and teachers. In S. N. Friel & G. W. Bright (Eds.), *Reflecting on our work: NSF teacher enhancement in K-6 mathematics* (pp. 179-186). Lanham, MD: University Press of America.

Curcio, F. R. (1987). Comprehension of mathematical relationships expressed in graphs. *Journal for Research in Mathematics Education, 18*, 382-393.

Curcio, F. R. (1989). *Developing graph comprehension: Elementary and middle school activities.* Reston, VA: National Council of Teachers of Mathematics.

Fennema, E., Franke, M. L., Carpenter, T. P., & Carey, D. A. (1993). Using children's mathematical knowledge in instruction. *American Educational Research Journal, 30*, 555-584.

Friel, S. N., & Bright, G. W. (1995). *Assessing students' understanding of graphs: Instruments and instructional module.* Chapel Hill, NC: University of North Carolina at Chapel Hill, University of North Carolina Mathematics and Science Education Network.

Friel, S. N., & Bright, G. W. (Eds.). (1997). *Reflecting on our work: NSF teacher enhancement in K-6 mathematics.* Lanham, MD: University Press of America.

Friel, S. N., Bright, G. W., Frierson, D., & Kader, G. D. (1997). A framework for assessing teachers' knowledge and students' learning of statistics (K-8). In I. Gal & J. B. Garfield (Eds.), *The assessment challenge in statistics education* (pp. 55-63). Amsterdam, The Netherlands: IOS Press.

Fuys, D., Geddes, D., & Tischler, R. (1988). The van Hiele model of thinking in geometry among adolescents: *Journal for Research in Mathematics Education Monograph, 3.*

Gal, I., & Garfield, J. B. (Eds.). (1997). *The assessment challenge in statistics education.* Amsterdam, The Netherlands: IOS Press.

Silver, E. A., & Smith, M. S. (1996). Building discourse communities in mathematics classrooms: A worthwhile but challenging journey. In P. C. Elliott & M. J. Kenney (Eds.), *Communicating in mathematics, K-12 and beyond: 1996 yearbook* (pp. 20-28). Reston, VA: National Council of Teachers of Mathematics.

McKnight, C. C. (1990). Critical evaluation of quantitative arguments. In G. Kulm (Ed.), *Assessing higher order thinking in mathematics* (pp. 169-185). Washington, DC: American Association for the Advancement of Science.

Vacc, N. N., & Bright, G. W. (in press). Elementary preservice teachers' changing beliefs and instructional use of children's mathematical thinking. *Journal for Research in Mathematics Education.*

Wainer, H. (1992, February/March). Understanding graphs and tables. *Educational Researcher, 21*(1), 14-23.

WGBH Boston. (1996). *Teaching math: A video library, 5-8* [video]. Boston, MA: WGBH Educational Foundation.

Changing Classroom Assessment Practices

William S. Bush
University of Kentucky

Roger and Jane are fifth-grade mathematics teachers in different school districts. Coincidentally, they are both in their tenth year of teaching. A look at their classroom assessment practices reveals many differences.

Roger uses the same types of classroom assessment that his mathematics teachers used when he was a student. Roger assigns homework almost every night directly from the textbook that dictates his lessons. The assignments generally include simple one-answer questions, multiple-choice items, sets of practice exercises, and an occasional word problem. Every Friday, he gives students a quiz on the topics covered that week. Typically, the quizzes include exercises and tasks like those on the homework, and students have 15 minutes to complete two or three problems. After each chapter, approximately every two weeks, Roger gives students a chapter test found in the teachers' edition of the textbook. It also includes exercises similar to those on homework assignments. Roger prepares students for the chapter test by assigning them review exercises the day before. For grades, homework is worth 20 percent, quizzes 20 percent, tests 50 percent, and class participation 10 percent.

Six years ago, Jane assessed her students as Roger does - homework, weekly quizzes, and chapter tests. Over the past few years, however, she has made some significant changes. She still gives homework, quizzes, and tests; however, the assessments now include a variety of tasks, including single-response questions, exercises, open-ended questions, and short problem-solving tasks. In addition,

Jane requires her students to keep a weekly mathematics journal in which they explain what they have learned or how they have felt about mathematics class during the week. Students also compile and submit a mathematics portfolio at the end of each 12-week grading period. The portfolio includes their best work from homework, quizzes, and tests, as well the results of classroom investigations or out-of-class group projects. Jane also uses checklists to assess her students' ability to solve problems and work in cooperative groups. Finally, Jane asks students to assess themselves and their peers on selected tasks and projects. She does not use letter grades. Instead, she provides parents a detailed checklist indicating how well students solve problems, reason and communicate with mathematics, and understand mathematics. She also includes descriptions of their feelings about doing mathematics, their ability to work in groups, and their ability to assess themselves.

Roger and Jane are not real teachers, but they represent composites of teachers I have spent time with over the past five years. I have been immersed in the educational reform effort in Kentucky and have spent much time working with teachers. The anchor for the Kentucky reform effort is a high-stakes state assessment system; teachers may receive cash rewards or possibly lose their jobs because of their students' performance on the state assessment. I have seen some teachers embrace the new forms of assessment (open-ended questions, performance events, and portfolios) with enthusiasm and vigor. I have watched other teachers try to change their assessment practices but fail. And I have watched still others resist the changes and fight to maintain their traditional forms of assessment. I have spent considerable time trying to understand why teachers like Jane make changes in assessment and why others do not.

I have struggled myself with issues of assessment as I have tried to change my own classroom assessment practices. I have reflected on my beliefs about learning and teaching and how they have affected my classroom practices. I have tried to understand what motivates me to make changes in my assessment practices. In this paper, I will share some of my analysis to date. I will identify some factors that I believe affects teachers' willingness or ability to change classroom assessment practices. I will also pose possible research questions that, if addressed, will provide further insight into understanding why some teachers embrace new forms of assessment and others resist changes.

Reasons for Change

Recent research on changes in classroom practices reveals many factors that encourage or discourage reform in teaching (Ball, 1988, 1993, in press; Brown & Borko, 1992; Clarke, 1997; Cooney, 1985; Lampert, 1987; Shifter, 1996; Thompson, 1984, 1992). For example, motivation, beliefs and values, dispositions, knowledge, and support strongly influence teaching practices. They also affect our willingness to explore and implement new approaches to assessment. Our beliefs about learning and teaching will affect what assessment approaches we choose. Our own understanding of mathematics affects the way we assess our students' understanding. Our willingness to take risks in the classroom will influence our ability to try new assessment ideas. Finally, access to assessment tools will have an impact on our ability to use a variety of assessments.

In this paper, I will delineate these factors for the sake of analysis and discussion. I fully understand that many factors intertwine and connect in complex ways to shape teaching practices. Factors that affect change in assessment practices can fall into two general categories: intrinsic (within us) and extrinsic (outside us and in our environment). At the end of the discussion about each factor, I also pose a set of research questions. These questions are intended to promote further discussion of factors that affect changes in assessment practices.

Intrinsic Factors

Motivation
As humans, motivation drives our actions. We change because we are motivated to change. On some occasions, motivation to change comes from within, or is intrinsic, for example, we change to portfolio assessment because we think that portfolios will provide us richer information about students. Some motivation comes from outside of us, or is extrinsic, for example, we change to portfolio assessment because our district or state requires us to use them.

Perhaps our assessment practices differ because we hold different intrinsic motivation. Recent literature on classroom assessment asserts that our primary reason for assessing students should be to find out what students know and can do mathematically (Stenmark, 1989, 1991; Clark, 1988). As teachers, we use this information to provide feedback to our students, to inform parents and others, and to evaluate our teaching. Perhaps some of us change our assessment practices because we are motivated to find out more about our students. Perhaps some of us are motivated by a desire to prepare our students for placement assessments like the SAT or ACT. Or, perhaps we are motivated to keep our instructional and assessment practices in line with our curriculum demands. In any case, motivation plays an important role in our propensity or reluctance to change practices.

Reflections on research. The powerful influence of intrinsic motivation on change in assessment practices makes it an important focus for research. Understanding what motivates us to change their assessment can help us learn to work with colleagues, administrators, and parents in making school or district changes. This information can be particularly useful for professional development providers and university faculty in their roles of helping teachers with assessment. Some important research questions related to intrinsic motivation might include:

- Why do teachers change their classroom assessment approaches?
- Why don't teachers change their classroom assessment approaches?
- How do changes in motivation relate to changes in assessment approaches?
- How does the type of motivation affect the quality and quantity of change in assessment approaches?
- How might professional development or preservice education address intrinsic motivation?
- What professional development approaches might be used to help teachers' understand and change, if necessary, their motivation for classroom assessment?

Beliefs and Values

Beliefs and values also influence our instructional practices (Schmidt & Buchman, 1983; Thompson, 1984, 1992). The decisions we make as teachers are influenced by what we believe and value. If we value computational proficiency, we have students practice computational exercises. If we value communication in mathematics, we ask students to write. What we value in mathematics teaching is reflected in how we assess student learning. Students learn very early in their schooling that assessment reflects what we, as teachers, think is important.

Our beliefs about the nature of mathematics also affect our assessment practices. If we view mathematics as a dynamic structure of relationships, we assess students' ability to understand relationships and make connections. If we view mathematics as a disconnected, rule-oriented discipline, we assess students' ability to master individual procedures. If we view mathematics as a tool for solving problems, we assess students' ability to solve problems and apply mathematics to new contexts.

We also differ in our beliefs about assessment types. For example, we might think that some types of assessments are more reliable or valid than others. We tend to trust the results of multiple-choice tests, and we question the results of portfolios. For example, some of us may believe that projects and portfolios are reliable ways to learn more about her students. Some of us, on the other hand, may not trust the results of open-ended questions, performance events, or portfolios; the information they provide seems far too subjective.

We also differ in our beliefs about the role of assessment. Some types of assessment provide very specific information about students while other types provide more general information. Some types of assessment are designed to compare students; other types are designed to assess individual growth. For example, some of us might use journals and portfolios because they provide the best information about students.

Reflections on research. The relationship between beliefs and teaching practices has been clearly established through research (Schmidt & Buchman, 1983; Thompson, 1984, 1992). We would expect the same relationships between beliefs and assessment approaches. Pertinent research questions related to beliefs, values, and assessment might include:

- How do beliefs about learning and teaching relate to beliefs about assessment?
- How do beliefs about learning, teaching, and assessment relate to assessment practices?
- Do instructional goals and practices align with assessment practices?
- Do changes in assessment approaches follow changes in beliefs about learning and teaching or vice versa?

- What assessments do teachers accept as reliable, valid, and trustworthy?
- What professional development models are most effective in changing beliefs and values, and ultimately assessment approaches?

Disposition

Sometimes emotions, affect, or personality influence our willingness or ability to make changes. Personality characteristics such as self-confidence, self-attribution, efficacy, and locus of control affect our willingness to take risks and attempt new ideas and approaches. Negative emotions like apathy, depression, discouragement, and anxiety can often block our ability to make changes. When we are stressed, we are not usually able to deal with change. On the other hand, when we are happy, joyful, enthusiastic, or confident, we tend to risk making changes. Factors like confidence and efficacy develop and persist over long periods of time, while apathy, depression, and anxiety are often short-term and tend to ebb and flow often.

Reflections on research. Previous studies on the dispositions of teachers have focused primarily on their effects on teaching and student learning (Aiken, 1970; Buhlman & Young, 1982; Schofield, 1981; Williams, 1988). Similar results are expected regarding their effects on classroom assessment practices. Some research questions related to dispositions and assessment practices include:

- What dispositions affect teachers' propensity to change assessment practices?
- To what extent do these dispositions affect teachers willingness or reluctance to change assessment practices?
- How can emotional and affective factors which deter changes in assessment practices be addressed through professional development or support systems?

Knowledge about Mathematics

Our understanding of mathematics and our mathematical ability has an impact on what and how we assess our students. I contend that we will not, or cannot, assess what we do not know ourselves. If we are not good problem solvers, we probably will not pose problems or assess problem-solving ability. If we are not good writers, we will probably not to ask our students to write or assess our students' understanding through writing. If we do not understand particular mathematical concepts or skills, we will not, or cannot, assess our student's understanding of those concepts or skills.

Reflections on research. If we, as teachers, are to assess many realms of mathematical knowledge and ability, we must be mathematically capable ourselves. The follow questions raise important issues concerning the impact of mathematical knowledge on assessment practices:

- Can teachers assess what they do not know?
- What types of mathematical knowledge do teachers need in order to assess students using a variety of tools?
- What kinds of preparation programs and professional development can provide teachers the mathematical knowledge base necessary to assess students competently?

Knowledge about Assessment

Knowledge about assessment can also affect our ability to change assessment practices. Knowledge about assessment can be considered at different levels. The levels of implementing new teaching approaches, developed by Loucks-Horsley, Harding, Arbuckle, Murray, Dubea, and Williams (1987), provide a way to categorize these levels of knowledge. The lowest level is awareness. For example, some of us may be aware of a greater variety of assessment approaches than others.

The next level is implementation. Learning to develop a reasonable balance between instruction and assessment is initially difficult. We must often give up something in our lessons in order to use other assessment approaches, especially when encouraging students to assess themselves and others. Making these shifts in practice requires an arsenal of time-saving and management strategies. Without these strategies, assessment can be cumbersome, awkward, and overwhelming to our students and us. For example, some of us might have learned to use a variety of assessment approaches effectively and efficiently through professional development. Some of us, on the other hand, may not have developed the skills to implement different assessments successfully.

The third level is understanding. At this level, we understand the role of assessment in our educational enterprise. We see how assessment fits our instructional goals and aligns with our beliefs. This level of understanding generally comes with considerable trial and error accompanied by reflection. For example, some of us may have developed a deeper understanding of the goals, power, and caveats of the assessment process. Through study and reflection, we come to understand how assessment fits into our vision and goals. Others of us may have not thought much about our assessment approaches and whether they align with our instructional goals.

Reflections on research. Research on teachers' implementation of changes reveals three levels of knowledge. Knowledge about assessment practices is worthy of further study. Research in this area might address the following questions:

- What types of knowledge do teachers hold about assessment?
- How does teachers' knowledge about assessment relate to their assessment practices?
- How do changes in levels of knowledge affect changes in assessment?
- How do levels of knowledge affect teachers' ability to change assessment practices?
- What professional development approaches are effective in building teachers' knowledge about assessment?

Extrinsic Factors

Factors in our teaching environment can also affect our willingness and ability to change assessment practices. The political, social, and financial forces in our world have powerful influences on what we do in our classrooms. In this section, I will focus on two broad, but important, extrinsic factors: motivation and support.

Motivation

Motivation can be extrinsic or intrinsic depending on the source. The sources of extrinsic motivation may be found at many levels: the classroom, school, district, community, state, or nation. Decisions and actions at all these levels can have a profound impact on our assessment decisions. For example, some of us change our assessment practices because of pressure from administrators or peers. Others refuse to change our practices because parents do not understand alternative approaches to assessment. The process of education has many stakeholders - students, parents, administrators, colleagues, experts, and community members. As teachers, we must consider the beliefs, values, and attitudes of these stakeholders. District, state, or national tests can provide extrinsic motivation for changing classroom assessment practices. We often alter our classroom assessment to prepare students for these tests.

Extrinsic motivation can lead to poor classroom assessment practices. For example, pressure from outside can often lead us away from our primary goals for classroom assessment - learning about students and assessing our teaching. Some Kentucky teachers with whom I have worked offer a good example of this influence. Five years ago, Kentucky implemented a high stakes accountability system based on its state assessment. Stories abound of teachers' having students practice answering open-ended questions *ad infinitum* just before state assessment days. Some Kentucky teachers require students to complete unrelated, disjointed mathematics tasks over weekends just to have entries for their mathematics portfolios. These practices differ considerably from those teachers who use open-ended questions and portfolios to find out what mathematics students know and can do.

Reflections on research. The questions regarding the impact of extrinsic motivation on changes in assessment practices are similar to those for intrinsic motivation:

- What external factors cause teachers to change their classroom assessment approaches?
- What external factors discourage teachers from changing their classroom assessment approaches?
- How do external factors affect the quality and type of assessment practices?
- What strategies might be used to help teachers address influences from outside sources?

Support

When implementing changes in our practices, we must have considerable support. This support may come in many forms: resources, time, and emotional. We must have a variety of resources to implement new types of classroom assessment. Ready access to collections of good tasks and questions, to relevant checklists and observation forms, and to samples of criteria and rubrics for scoring assessment will enhance the likelihood that our attempts at change will be successful. For example, some of us change our assessment practices simply because we had access to a resource room containing assessment types and literature about assessment. Others of us have not changed because access to tasks and tools are limited.

Along with the support of physical materials, we also need time to change assessment practices. Current school structures and schedules do not support change. We have little time to think about alternative assessment, to read about assessment or to gather assessment materials. Alternative assessments like portfolios, interviews, and performance events require more time to implement and score. Some of us have been able to change our assessment practices because principals supported the staff by structuring the school day to allow common planning time for teachers.

Strong emotional support systems are necessary to deal with the stress that comes from change. We need support from colleagues, administrators, parents, and students to implement innovative practices. The emotional support provided by working with colleagues on issues of assessment is invaluable. For example, some of us change our assessment practices because other teachers in our school are also changing their practices. Some of us have developed a support system by attending conferences and workshops about assessment.

Emotional support can come from outside schools and districts. For example, the Kentucky Department of Education employs exceptional teachers to work with other teachers in their region in improving assessment strategies. These exceptional teachers not only support teachers in developing a knowledge base but also provide much emotional support. Policies at the state and national levels can have a strong impact of the nature and quality of support that teachers receive.

Reflections on research. Resources and support are necessary ingredients in the change process. The following questions raise important issues about the relationship among support and change in assessment practices:

- What resources do teachers need to change their assessment practices?
- How does availability of resources affect teachers willingness or ability to change assessment practices?
- How do teachers use assessment resources provided to them?
- What support is needed with resources to assist teachers in making changes in their assessment practices?
- What kinds of support do teachers need to make changes in classroom assessment practices?
- What levels of support (school, district, state, nation) have the most impact on changes in assessment practices?

- What models of support systems are effective in helping teachers change their classroom assessment practices?
- What are the costs, in time and money, in developing effective support systems?

The Interplay of Factors

As mentioned earlier, these factors do not reside in isolation within our teaching world. They intertwine in complex ways to enable or deter us from making changes in our practices. Consider the following vignette:

> *Martha has been thinking about her assessment practices for a while. She is not satisfied with the evidence of learning she has accumulated about her students. She begins talking to other teachers, and they too have been concerned about their assessment practices. They approach their principal, and she agrees to support an all-day workshop for teachers in the school. An expert in classroom assessment comes to the school and helps the teachers understand the roles of assessment. Teachers also learn about alternative assessment tools and strategies. The teachers decide to try mathematics portfolios. The principal provides travel funds for the interested teachers to attend workshops about using portfolios. The school librarian orders several books about writing in mathematics for the school's professional library. The teachers meet monthly for several months as they try to implement mathematics portfolios.*

This scenario shows both intrinsic and extrinsic factors at work. The intrinsic factors include motivation (more evidence about student learning), knowledge (new tools and approaches), resources (professional library), time (attend workshops and conferences), and emotional support (monthly meetings).

Reflections on research. Will the teachers in this scenario succeed in their attempt to change their assessment practices? Maybe, some important factors have been addressed in this situation. Other factors, however, are not addressed. Do the portfolios align with their beliefs and values about assessment and instructions? Will the teachers develop the expertise to implement a viable portfolio system? Will parents support mathematics portfolios? Will students resist doing portfolios?

Will teachers have time to score portfolios? Will they continue to receive support from administrators after initial attempts? Our chance of success in changing our classroom assessment practices depends on the strength of both intrinsic and extrinsic factors. The following questions address how factors might interact in affecting change:

- What relationships exist among intrinsic and extrinsic factors?
- Are some factors more likely to effect change practices than other factors?
- Is change developmental? Do teachers tend to go through stages in changing their practices?
- What factors tend to elicit change that produces sustained practices over long periods of time?
- What factors are necessary to produced sustained changes in practice?
- What types of professional development are effective in addressing different factors?

Closing Remarks

Changing classroom assessment practices is not easy; many factors affect our ability to make substantial changes. My attempts at changing my practice and my work with teachers as they have tried to change their practices have caused me to think about important factors. Understanding what enables us to make changes and what keeps us from making changes allows us to search for solutions and support. What kinds of professional development are needed? How should assessment be addressed in teacher preparation programs? How might schools and districts support teachers? Many more questions about changing classroom assessment practices remain. Solutions are available through continued analyses and discussions of the issues.

References

Aiken, L. (1970). Attitudes toward mathematics. *Review of Educational Research*, *40*, 551-596.

Ball, D. L. (1988). Unlearning to teach mathematics. *For the Learning of Mathematics*, *8*, 40-48.

Ball, D. L. (1993). With an eye on the horizon: Dilemmas of teaching elementary school mathematics. *Elementary School Journal*, *93*(40), 373-397.

Ball, D. L. (in press). Teaching learning and the mathematics reforms: What do we think we know and what do we need to learn? *Phi Delta Kappan.*

Brown, C. A., & Borko, H. (1992). Becoming a mathematics teacher. In D. A. Grouws, (Ed.), *Handbook of research on mathematics teaching and learning* (pp. 209-239). New York, NY: Macmillan.

Buhlman, B. J., & Young, D. M. (1982). On the transmission of mathematics anxiety. *Arithmetic Teacher, 30*(3), 55-56.

Clarke, D. (1988). *Assessment alternatives in mathematics: MCTP professional development package.* Melbourne, Australia: Curriculum Corporation.

Clarke, D. M. (1997). The changing role of the mathematics teacher. *Journal for Research in Mathematics Education, 28,* 278 -306.

Cooney, T. J. (1985). A beginning teachers' view of problem solving. *Journal for Research in Mathematics Education, 16,* 324-336.

Lampert, M. (1987). Mathematics teaching in schools: Imagining an ideal that is possible. In *The teacher of mathematics: Issues for today and tomorrow* (pp. 37-42). Washington, DC: National Academy Press.

Loucks-Horsley, S., Harding, C., Arbuckle, M., Murray, L, Dubea, C., & Williams, M. (1987). *Continuing to learn: A guidebook for teacher development.* Andover, MA: Regional Laboratory for Educational Improvement of the Northeast and Islands, & Oxford, OH: National Staff Development Council.

Schmidt, W., & Buchman, M. (1983). Six teachers' beliefs and attitudes and their curricular time allocations. *The Elementary School Journal, 84*(2), 162-171.

Schofield, H. L. (1981). Teacher effects on cognitive and affective pupil outcomes in elementary school mathematics. *Journal of Educational Psychology, 73,* 462-471.

Shifter, D. (Ed.). (1996). *What's happening in math class? Envisioning new practices through teacher narratives.* New York, NY: Teachers College Press.

Stenmark, J. K. (1989). *Assessment alternatives in mathematics: An overview of assessment techniques that promote learning.* Berkeley, CA: Regents, University of California.

Stenmark, J. K. (Ed.). (1991). *Mathematics assessment: Myths, models, good questions, and practical suggestions.* Reston, VA: National Council of Teachers of Mathematics.

Thompson, A. G. (1984). The relationship of teachers' conceptions of mathematics and mathematics teaching to instructional practice. *Educational Studies in Mathematics, 15,* 105-127.

Thompson, A. G. (1992). Teachers' beliefs and conceptions: A synthesis of research. In D. A. Grouws, (Ed.), *Handbook of research on mathematics teaching and learning* (pp. 127-146). New York, NY: Macmillan.

Williams, W. V. (1988). Answers to questions about mathematics anxiety. *School Science and Mathematics, 88,* 95-104.

Computers in Mathematics Education Assessment

Douglas H. Clements
State University of New York at Buffalo

My concept of myself as a student has suffered somewhat since I began (secretly) taking my sixth-grader's science tests. I am fine on the mathematics, mind you, but the biological terms decimate my scores. How would you do, today, on a range of public school assessments? What does that mean?

Nature of Assessment in Mathematics

In far too many cases, we are testing, and therefore teaching, the wrong things. The long history of laments regarding inert knowledge will continue unabated unless we transform mathematics assessment. Reductionist pieces of information are in part the result of national, standardized, instruments that test only that at the intersection of myriad rich and shallow curricula (Lipson, Faletti, & Martinez, 1990). Mathematics innovation cannot grow in such environments.

Instead, we need additional forms of assessment, including those called "authentic," "performance," or "direct." Such forms are based on tasks that range from simple but meaningful to complex, integrated, and challenging. Most of these tasks should, to varying degrees, require reflection, emphasize important concepts, and demand persistence. School systems, especially those that score low, will find a way to "reduce the level" (van Hiele, 1986) of instruction to improve scores on any other type of assessment of "better" or "higher-order" mathematics (Baker, 1990; Frederiksen & Collins, 1989).

The purpose of this paper is to suggest ways that computers can help promote such assessment and therefore contribute to reform and innovation in mathematics education.

Computer-enhanced Assessment

Computers may enhance assessment in a variety of ways. For example, typical feedback from printed tests is the delayed presentation of a score and rank. Computers can provide more timely and elaborated feedback. For example, if a problem involved calculating paint to cover a wall, the computer might show the difference between the given area and the area covered by the amount of paint the student calculated (Lipson et al., 1990). In such ways, computer-based assessment can be more instructionally powerful. There are, in addition, many ways computers can significantly alter assessment.

Adaptive Testing
A relatively simple, but still underutilized, form of computer-enhanced testing is adaptive testing (Lesh, 1990; McKinley & Reckase, 1980). During testing, the computer selects items that are deemed appropriate for the student based on her or his previous responses. If the student's performance indicates that she or he will not be able to handle certain types of times, these are not presented. Computer testing also allows the efficient use of open-ended items and new questioning sequences rather than the restricted multiple-choice format. Multiple versions allow multiple administrations and tracking progress rather than only summarizing outcomes. Errors of measurement are minimized, testing time shortened (from 1/2 to 1/3 of the time for conventional assessment), frustration reduced, reliability improved, and total test information gathered increased, including profiles of strengths and weaknesses rather than unidimensional scores. This type of assessment, then, takes less time away from, and contributes more to, classroom instruction.

Generally, research has supported the efficacy of computer diagnosis and remediation of arithmetical errors. Future software developments should promote the transition from multiple-choice formats with a single correct choice to constructed-response questions and open-ended questions and problems that test higher-level knowledge and skill. Eventually, such software might present many different kinds of problems (with different contexts generated to match the interests of each student), and track and interpret not only answers, but intermediate steps in solution attempts (Lipson et al., 1990).

In a similar vein, adaptive instructional systems perform an initial diagnostic assessment of each student (or skip this time-consuming process if prior information indicates minimal existing knowledge).

Then assessment is continually updated and new prescriptions are made as a student works. The amount of instruction given is based on the student's progress toward the mastery of the objectives, which reduces off-task time (Tennyson, Christensen, & Park, 1984).

Systems such as this can help establish connections among related ideas, both within mathematics and across other subject matter areas. Systems that generate an array of interdisciplinary problems and track multiple concepts and processes may facilitate the development of comprehensive interdisciplinary instruction (Lipson et al., 1990).

Increasingly Intelligent Assessments

As systems become more intelligent, valid micro-assessments become achievable (Snow, 1989). Enhanced with a model of the learner (Frederiksen & Collins, 1989), computers can make more finely tailored assessment and instructional decisions. For example, Lesgold's (1988) tutor for troubleshooting electrical circuits continually updates its student model in relation to a previously specified goal structure and designs the next task to be presented based on a list of constraints built up from previous assessment results. This list contains not only the state of learning progress in the domain but also constraints representing other relevant initial states of learners, such as initial mathematical ability. For example, the following decision might be made: A student is ready to move to the next specified level of complexity in resistor networks, but is weak in arithmetic skill, so the voltage computations will be kept mathematically simple. In this way, the tutor circumvents some aptitude weaknesses while working on other transitions.

For mathematics, systems should begin to be developed in domains that research has explored in detail (Lipson et al., 1990). For example, Marshall (1990) has applied schema theories to build computer-based assessments of solving arithmetic story problems. A variety of item types provided distinct information that judged the level of understanding demonstrated by different responses.

Such systems eventually may be able to provide adaptive assessment and instruction, and also analyze and interpret a student's performance beyond correctness; make sophisticated analyses of multiple step problems; incorporate techniques of dynamic assessment; allow assessment of higher-order thinking, circumventing the need for beginners to automatize skills; maintain a dynamic model and representation of the student's knowledge and skill; access a data base of ιtics relevant to the curriculum, including applications outside ᴏ.   ϲhool context; provide progress maps depicting the student's learning over time, suggesting possible areas for additional study, and mapping out the "landscape" ahead of the knowledge to be gained, including possible points of difficulty; and make instructional suggestions (Lajoie, 1990; Lipson et al., 1990).

Simultaneous Group and Individual Assessment

Though research is limited, there is also significant potential in new systems that allow multiple students to communicate with a teacher. For example, each member of a class might enter their response to a problem, which would all be instantly transmitted to the teacher. The teacher can examine, display, and discuss various solutions with the class, monitor the responses of the class as a whole, of groups, or of individuals, and store information at each of these aggregate levels. Students might use keypads or calculators to enter their solutions.

"Doing Mathematics"

To assess student's ability to fully "do mathematics," the state of the art still requires subjective assessments that include performance on extended tasks (Frederiksen & Collins, 1989). Computer traces of student work processes can help document such performance. However, the computer must be integrated into a larger assessment effort.

Further, the computer can make additional contributions, many intimated previously. For example, they can be used to create simulations and microworlds that are a rich source of problems. Work with Logo and other environments has provided "windows to the mind" for many teachers and researchers (Chazan, 1991; Clements & Meredith, 1993; Weir, 1987). Weir, for example, showed that children's first choices of numbers as input to the Logo "forward" command revealed much about their number sense.

Teachers similarly learn much about their students exploring mathematics concepts, including fractions, with *Turtle Math* (Clements & Meredith, 1994). In one case, students had completed a week of instruction on fractions, but when encountering fractions on the computer, they relied on instinct (Sarama, 1995). For example, Linda was playing Berry Good Meal, an activity in which she was to enter a forward command that enabled the turtle to eat berries placed on a line segment somewhere from 0 to 1 (in *Turtle Math*, the turtle's step size can be changed; in this activity, it was 200 pixels. Linda guessed $\frac{2}{3}$; when the turtle did not go far enough, she wanted to try $\frac{2}{4}$. When that did not work, she said, "Of course," and mentioned that $\frac{2}{4}$ is the same as $\frac{1}{2}$. Though she originally was trying to change the denominator by rote to make the fraction smaller, the feedback from the turtle encouraged her to reflect on what she had entered and make quantitative sense of the fractions. The teacher gained valuable knowledge about Linda's ideas and strategies with fractions. This teacher also observed that computer environments encouraged students to try their own algorithms and thus become mathematical empiricists, rather than merely receivers of knowledge from authorities. When trying to play Berry Good Meal in a way in which the input to the fd command was a sum, two girls, Carrie and Heather, tried $\frac{1}{3} + \frac{1}{3}$ for what they thought was $\frac{1}{3}$, with Heather reasoning $\frac{1}{3} + \frac{1}{3}$ is $\frac{2}{6}$ (adding numerator and denominator) which reduces to $\frac{1}{3}$. Carrie disagreed, but Heather was insistent that her logic was correct, so they tried it. Carrie used the feedback from the turtle as proof that Heather was wrong and Heather worked to reconcile her original idea with that feedback (Sarama, 1995). Such insights into children's epistemological actions is essential for the teacher attempting to implement mathematics education reform.

Recommendations

1. Change the basic approach to assessment.
   - De-emphasize reductionist, indirect testing in favor of direct assessment so that systems cannot reduce the level of educational goals.
   - Eliminate paper-and-pencil tests from the earliest years.
   - Eliminate unidimensional, passive indicators in favor of assessment as an integral part of complex, dynamic, self-regulating organic system.
   - Eliminate most separate assessments; most assessment should occur within instructional activities.

2. Recognize that innovation in mathematics education can be substantially enhanced through technology. Use computers for
   - enhancing feedback
   - adaptive testing
   - increasingly intelligent assessment
   - simultaneous group and individual assessment
   - providing environment for "doing mathematics"
3. Consider the need to improve teacher education on assessment, the use of computers, and their synthesis.

References

Baker, E. L. (1990). Developing comprehensive assessments of higher order thinking. In G. Kulm (Ed.), *Assessing higher order thinking in mathematics* (pp. 7-20). Washington, DC: American Association for the Advancement of Science.

Chazan, D. (1991). *Research and classroom assessment of students' versifying, conjecturing, and generalizing in geometry.* Unpublished manuscript, Michigan State University, East Lansing, MI.

Clements, D. H., & Meredith, J. S. (1993). Research on Logo: Effects and efficacy. *Journal of Computing in Childhood Education, 4,* 263-290.

Clements, D. H., & Meredith, J. S. (1994). *Turtle math* [Computer program]. Montreal, Quebec: Logo Computer Systems, Inc. (LCSI).

Frederiksen, J. R., & Collins, A. (1989). A systems approach to educational testing. *Educational Researcher, 18*(9), 27-32.

Lajoie, S. P. (1990, April). *Computer environments as cognitive tools for enhancing mental models.* Paper presented at the meeting of the American Educational Research Association, Boston, MA.

Lesgold, A. (1988). Toward a theory of curriculum for use in designing intelligent instructional systems. In H. Mandl & A. Lesgold (Eds.), *Learning issues for intelligent instructional systems* (pp. 114-137). New York: Springer.

Lesh, R. (1990). Computer-based assessment of higher order understandings and processes in elementary mathematics. In G. Kulm (Ed.), *Assessing higher order thinking in mathematics* (pp. 81-110). Washington, DC: American Association for the Advancement of Science.

Lipson, J. I., Faletti, J., & Martinez, M. E. (1990). Advances in computer-based mathematics assessment. In G. Kulm (Ed.), *Assessing higher order thinking in mathematics* (pp. 121-134). Washington, DC: American Association for the Advancement of Science.

Marshall, S. P. (1990). The assessment of schema knowledge for arithmetic story problems: A cognitive science perspective. In G. Kulm (Ed.), *Assessing higher order thinking in mathematics* (pp. 155-168). Washington, DC: American Association for the Advancement of Science.

McKinley, R. L., & Reckase, M. D. (1980). Computer applications to ability testing. *AEDS Journal, 13,* 193-203.

Sarama, J. (1995). *Redesigning Logo: The turtle metaphor in mathematics education.* Unpublished doctoral dissertation, State University of New York at Buffalo, Buffalo, NY.

Snow, R. E. (1989). Toward assessment of cognitive and conative structures in learning. *Educational Researcher, 18*(9), 8-14.

Tennyson, R. D., Christensen, D. L., & Park, S. I. (1984). The Minnesota adaptive instructional system: An intelligent CBI system. *Journal of Computer-Based Instruction, 11,* 2-13.

van Hiele, P. M. (1986). *Structure and insight.* Orlando, FL: Academic Press.

Weir, S. (1987). *Cultivating minds: A Logo casebook.* New York, NY: Harper & Row.

A Through the Lens Look at Moments in Classroom Assessment

Francis (Skip) Fennell
Western Maryland College
and
Mathematical Sciences Education Board

Classroom assessment serves to advise and monitor instruction. For far too many beginning teachers, classroom assessment is an instructional afterthought, something that is done to help determine grades or judge progress. Such formative assessment strategies must become integral to instructional planning, teaching, and summative assessment. At a time of reform-based mathematics curriculum initiatives, the acquisition of a repertoire of classroom assessment strategies linked to instruction is critical to the success of such initiatives.

Early Views - A Narrow Lens Perspective

Preservice programs in teacher education do not prepare beginning teachers to fully comprehend the linkage between instruction and the consistent use of a variety of classroom assessment practices. Beginning teachers, like novices in any field, do not have the experience and depth of background to fully understand such a link. This is a strong, but accurate statement. While the use of classroom assessment practices may be an integral component of assessment-related or mathematics education courses and related practicums and experiences, the task of assuming the role of classroom teacher is, at least in the beginning, to initiate and execute an instructional model which values the varied and consistent use of classroom assessment. Picture the first year teacher - anywhere. Somewhere after getting a classroom ready for students, and the seemingly minute-by-minute demands of classroom management and the time invested in just figuring out where things are and how to get routine classroom (copying, materials) and school issues (attendance, lunch, scheduling) completed, is the real need to understand the mathematics curriculum. In most schools, the daily "curriculum" is driven by a textbook, which may be supporting the state's frameworks, and/or the local school district's curriculum. The point of this "dash of reality" is that at the beginning stages of the teacher's career she is not prepared to see assessment as anything other than periodic "bumps" or moments in curricular delivery. This initial view of classroom assessment is seen through a narrow lens which views assessment as a periodic formal event to measure classroom achievement, that is, what do the students know? How did they do on the measurement unit?, etc., and is further colored by the need to generate progress indicators for report cards (Figure 1).

Moments	Assessment Purpose
Early Views: A Narrow Lens Perspective	
Tests, Quizzes	Classroom achievement
Expanded Views: A Wider Lens Perspective	
Homework assignments	Individual progress
Classroom observation	Class understanding
Individual students observation	Individual student understanding
Classroom questioning and discussion	Student and class understanding
Journals, logs	Student understanding and disposition
Experienced Views: Linking Assessment to Instruction	
Diagnostic interviews	Individual student understanding
Portfolios	Individual student achievement, understanding, and disposition
Performance tasks	Class and individual achievement and understanding
Norm referenced achievement tests	Class and individual achievement
Norm referenced aptitude tests	Class and individual general aptitude

Figure 1. Assessment Moments and Purposes in the Classroom

An Expanded View - A Wider Lens Perspective

As teachers gain experience and with it the opportunity to more fully understood the mathematics curriculum, they begin to adapt their instruction to meet specific curricular needs of their students, and recognize the critical linkage between instruction and assessment. With such understanding and experience their view of classroom assessment widens to recognize the informal assessment values to be derived from classroom observation and discussions, seat work, homework, and questioning. Student responses to classroom questioning, homework assignments, and group projects take on a new perspective. The mistakes and insights of students become more than mere moments in a long day, rather they become pieces in the intricate puzzle of classroom and student understanding.

Additionally, as teachers begin to see how closely their daily instruction is linked to classroom and curricular assessment, they will begin to acknowledge, albeit informally, a critical aspect of assessment, content validity. That is, they will be able to make sure that their assessment matches important mathematics learning goals. As teachers develop a greater understanding of such goals they are likely to assess the prior knowledge of their students by discussing lesson or unit goals with their students. They will also recognize the value of interviewing students to diagnose the depth of understandings. This view of using informal assessment to monitor and inform instruction is way beyond just noting the mistakes students make in classroom assignments.

An Experienced View - Linking Assessment to Instruction

Teachers with the widest view or lens for linking classroom instruction and achievement recognize the need to gather and interpret information about student achievement, aptitude, and disposition. Such information or moments in assessment have now been broadened to include observing student performance, the collection and analysis of work samples, math journals, student interviews, questioning, reviewing student group projects, portfolios, student tests and quizzes, and other opportunities. The quality of such evidence is truly in the eyes of the beholder! A work sample may be as revealing as a unit test, and a five-to-ten minute student interview may exhibit student achievement and disposition in a far more revealing way than the weekly class quiz. Such formative assessment is the on-going link to day-to-day instruction. As teachers come to recognize and value assessment from the broadest (or widest) perspective, assessment will become an integral part of the planning of every lesson. Teachers must value assessment and be able to think about assessment as they plan, implement, and formally assess a lesson, unit, or curriculum. It may be safe to assume that the more teachers understand and value assessment, the more varied the use of assessment in their repertoire.

In *Standards for Teacher Competence in Educational Assessment of Students* (National Research Center on Assessment, Evaluation, and Testing, 1990, p. 31-32) it is suggested that teachers need to develop competence and skill in the following areas:

- Standard 1: Choosing assessment methods appropriate for instructional decisions.
- Standard 2: Developing assessment methods appropriate for instructional decisions.
- Standard 3: Administering, scoring, and interpreting the results of both externally-produced and teacher-produced assessment methods.
- Standard 4: Using assessment results when making decisions about individual students, planning teaching, developing curriculum and school improvement.
- Standard 5: Developing valid pupil grading procedures which use pupil assessments.
- Standard 6: Communicating assessment results to students, parents, other lay audiences, and other educators.
- Standard 7: Recognizing unethical, illegal, and otherwise inappropriate assessment methods and uses of assessment information.

The standards presented above provide benchmarks for teacher background relative to assessment. However, they fall short in emphasizing the need, impact, and influence of classroom assessment. At a time when high stakes assessment (state testing programs, competency examinations, etc.) is driving instruction in many classrooms throughout the country, the linkage between classroom assessment and instruction is critical. It will only be when such links are made that classroom teachers will see how their daily work can influence high stakes testing events. As teachers begin to develop a repertoire of assessment strategies which include assessment moments as informal as questioning and observation and as engaging as group performance assessment tasks, they will draw external assessment events closer to their daily work.

As noted, teachers, for the most part, lack background, experience, and confidence in dealing with assessment. Prospective and inservice teachers need experience planning, implementing, and interpreting a variety of assessment strategies - from observations through performance assessment. Staff development opportunities are needed to assist in "widening the assessment lens." A recent survey (Fennell, 1997) of a graduate student population (all currently teaching) indicated limited opportunity, at the school level, to discuss the use of student assessment in making instructional decisions. Such dialogue is critical if we seek to increase and broaden teacher skills and confidence in assessment.

The *Standards for Teacher Competence in Educational Assessment of Students* cited earlier indicates that feedback to others - teachers, students, parents, is an important purpose of assessment. Assessment assists in providing feedback on student learning, informing and guiding instruction, and communicating learning experiences. While providing feedback on varied alternative assessments (portfolios, performance tasks, group projects) is not an easy task, parents will support such initiatives (Shepard & Bliem, 1995). Once again, with experience feedback becomes more varied and more frequent. While the report card remains as a historic school progress communicator, such feedback may also consist of periodic notes or letters home to report on progress and curricular experiences, and parent-teacher and parent-student-teacher conferences.

Portfolio assessment provides an excellent opportunity for parents, students, and teachers to reflect on the growth made in various aspects of mathematics learning - from daily work samples to math journals to weekly quizzes and tests. The portfolio has the potential to provide a variety of assessment "snapshots" for all those concerned.

Much of the current discussion, writing, and initiative in assessment today somehow relate to alternative assessment, that is, thinking about assessment in a way other than conventional, traditional models. It is fair to assume that some assessment procedure that is an alternative to the multiple choice, norm-referenced, standardized achievement test, has made its way into many, if not most, classrooms in this country. Performance assessment has become the most popular of the varied attempts of alternative assessment. The popularity of performance assessment can be traced to the reform-based initiatives (National Council of Teachers of Mathematics, 1989, 1995; National Academy of Science, 1989) in mathematics and other subjects. Most models of performance assessment has students responding to tasks which emphasize important mathematics concepts and skills. Such activities focus on the demonstration of higher order thinking and usually involve a written response. In some states student and/or school responses to performance assessment tasks has truly become high stakes testing. Schools and students are ranked, rated, and sorted based on such responses. While such accountability heightens student responsibility, in many cases, it may also contribute to an over-emphasis on student or school scores with less of an emphasis on curricular change and/or student or school improvement. Given the current discussion and planning regarding the President's eighth grade mathematics test and other national benchmarks, including the Third International Mathematics and Science Study (TIMSS), such concerns must be carefully monitored. The ultimate goal of any assessment initiative is to improve instruction and learning opportunities for children. As educators move from the narrow lens perspective of the apprentice teacher to the experienced vantage point of the wider, more broad based use of varied assessment strategies the boundary between classroom assessment, teaching and high stakes alternative assessment will be less defined.

References

Fennell, F. (1997, March). *Assessment survey - WMC/MCPS Project.* Wesminster, MD: Western Maryland College.

National Academy of Science. (1989). *Everybody counts.* Washington, DC: Author.

National Council of Teachers of Mathematics. (1989). *Curriculum and evaluation Standards for school mathematics.* Reston, VA: Author.

National Council of Teachers of Mathematics. (1995). *Assessment standards for school mathematics.* Reston, VA: Author.

National Research Center on Assessment, Evaluation, and Testing. (1990). *Standards for teacher competence in educational assessment of students.* Berkeley, CA: University of California.

Shepard, L. A., & Bliem, C. L. (1995). Parents thinking about standardized tests and performance assessments. *Educational Researcher, 24*(8), 25-32.

Classroom Assessment

Tery Gunter
Durham, NC, Public Schools

Good classroom instruction is based on assessment that is designed to inform instructional decisions. Master teachers continuously assess their students' understanding of subject matter. Teachers use the information gained through assessment to make instructional decisions. Information gained through quality assessment provides the teacher with the vehicle to enhance students' learning.

Assessment is the process of gathering evidence about a student's understanding of a subject such as mathematics. Teachers use several methods of gathering this evidence. One method is informal. This type of assessment occurs throughout the day when the class is discussing a topic and a child's verbal response reveals that child's poor understanding of a concept or his in-depth understanding. The following example illustrates how an informal observation during math class can inform the teacher about a child's understanding of telling time. The teacher explains to her first grade students how to tell time and has them set their paper clocks to 2:00, 3:00, 7:00, etc. As she walks around the room assessing how the students are doing on this skill one student stops her and says, "Look, I've set my clock at 3:18. That's the time on our class clock." The teacher realizes that this student's skill is probably beyond the first grade standard. She should work with this child further and assess his understanding of telling time. Informal assessment gives the teacher insight into a child's understanding. Observations made during this type of assessment should usually be followed by a more thorough assessment of the child's understanding.

Another method of assessment is individual interview. Teachers gain a lot of understanding about a child's learning through talking with him and asking questions about his knowledge. This assessment is different from the informal assessment in that it is planned. The teacher sets aside a time to evaluate a student. An example of this type of assessment is talking to a student about regrouping. The teacher could write three problems such as the following on a sheet of paper:

$$
\begin{array}{ccc}
5 & 76 & 82 \\
+\,17 & +\,22 & +\,79
\end{array}
$$

The teacher shows the student the first problem and asks him to solve it. There are various manipulatives on the table for the student to use to solve the problem. As he solves the problem the teacher has him talk to her about what he is doing. She takes notes. Next she asks him to solve the second problem and then the third. If while solving the first problem the teacher realizes that he does not understand regrouping, she can give him a simpler problem and assess where this child's understanding is. She then knows what to do to help him progress in regrouping. Anecdotal notes made during these interviews are kept in the child's portfolio and referred to when the teacher evaluates his understanding of the concept. The individual interview is very time consuming but provides the teacher with a much clearer picture of a student's understanding and what should be the next step for that student.

Writing in the mathematics classroom is a third important assessment tool which allows the teacher to "see" a student's thought process. As students write about how they solved a problem the teacher is able to better understand a child's thinking. Writing about how to solve a problem helps to develop a child's mathematical understanding and also links math to language arts. Teachers can learn a lot about a child's understanding of a mathematical concept by his written explanation. A written problem could be given to second graders; e.g., "I have 24 cookies. Is that an even or odd number of cookies? Explain your answer." With the student's written explanation, which could be in words or drawings, the teacher has a clearer understanding of the thinking about odd and even numbers. When students write about their solutions, the teacher has a window into her students' thinking.

In order to successfully teach children mathematics and assess their understanding teachers must keep three key ideas in mind. Teachers must know the mathematics curriculum that is expected to be taught at their grade level. Teachers should have a clear understanding of how the curriculum objectives will be assessed. After establishing clear curriculum goals and how they will be assessed, teachers must also establish the standard by which the work will be evaluated.

Communicating the objectives of the mathematics curriculum and how it will be evaluated to the students allows for greater student success. Students need and want to know what they will be learning and the standard expected of them. When an assignment is given to students the teacher should explain the expectation. A rubric is one way of communicating the standard. Rubrics are descriptions of performance expectations linked to a scale for evaluating the work. When the levels of proficiency are established and explained to the students, the students know the learning targets and what constitutes quality work. Rubrics help the teacher monitor student progress, assess the effectiveness of her instruction, and make summative statements about students' growth and development.

The information gathered about students' understanding is critical in making instructional decisions. Finding ways to record the information is also critical. In order to successfully assess student learning teachers have to develop record keeping strategies. Teachers need to find a method that works for them. If teachers are assessing students throughout the school day they are gathering a wealth of information about each child's learning. Keeping up with the information can be overwhelming if the teacher does not have a plan for record keeping. Teachers need to try several strategies and to develop their own system for recording, organizing and summarizing the information they gather on their students.

Record keeping methods that work for many teachers include student portfolios, checklists, post-it notes, and video recordings. If a portfolio or file is kept for each student, dated anecdotal notes and student work can be placed in that folder. Teachers must find a format for note taking that works for them. Writing observations on post-it notes and attaching them to student papers works for some teachers. Focusing on a few children at a time and having a clear understanding of the objective and the standard for that objective makes record keeping more successful. Sharing ideas, successes and failures, with other colleagues allows teachers to develop new record keeping methods.

Classroom assessment must be on-going and part of daily instruction. As teachers focus on classroom assessment as a means of making appropriate instructional decisions and of promoting student achievement, they find that they know more about their students than they ever have and that student learning increases.

Classroom Assessment: A Working Paper

Audrey Jackson
Fenton, MO, Schools

Assessment begins with a set of goals or expectations. These goals and expectations should be orchestrated to improve teaching and learning. If classroom assessment is to provide teachers with information to assist in curriculum design and decision making, teachers need to understand the discipline. As we move to the next millennium, educators continue to debate the "how" of classroom assessment.

Mathematics education has expanded to include communicating ideas, connecting ideas, and reflecting on ideas. It is essential for teachers and supervisors of mathematics to make connections and provide students with a meaningful learning environment. The National Academy of Science (1989) has reminded us to assess the development of students' mathematical power, use a mixture of means to reflect the mathematics curricula, and ensure that test measures what is of value.

When we mentor teachers (either new or veteran), we are addressing how to instruct, manage, and evaluate effective teaching and learning. In doing so, we must address assessment, since assessment is linked directly to instruction. Assessment is an essential part of teaching, and every effort must be made to ensure that whatever is measured is the result of what is taught. But the question is "How?"

Assessment, like learning, is an on-going process -- a process which includes formal and informal evaluating. Instruction in mathematics classes should serve as a catalyst which guides children through a continuum of creating knowledge, building skills, and developing dispositions. This is the same continuum where a child should read, write, think, and investigate numerous methodologies for a given concept.

As children discover, question, and interact, teachers should be observing work habits, listening to discussions, and assisting through questioning. It is essential to observe students' abilities to organize their work, develop their interests, and communicate with classmates. The development of skills and confidence in the use of numbers and in understanding and applying mathematical concepts is the key for mathematical power. Keeping a classroom journal of what you see and hear, including students explanations and responses, will prove to be invaluable when reporting progress.

Reports about students' progress should be based on information about everything that has happened in the classroom. This information evolves over time and comes from many sources:

- projects, which reveal oral and written skills
- portfolios, which provide insight about conceptual understanding, mathematical thinking, and expression of ideas
- group work, which reveal leadership and cooperation skills
- homework, which gives feedback for students, teachers and parents
- student self-assessment, which shows students' beliefs and self-perception
- collaborative student-teacher rating scale, which can help students create realistic expectations

Wiggins (1989) believes that good teaching is inseparable from good assessing. The observation of students and their performance is the principal tool of assessment. Open-ended problems allow students to respond in writing or with diagrams and pictures. Giving students time to think, and think again, provides a foundation for discovery and understanding.

It is more rewarding to engage your students in the learning process than to teach with the premise that all children learn at the same rate at the same time. It is more beneficial for students and teachers alike to promote higher level thinking skills and creative strategies for solving problems than to drill and practice skills. After all, it is more critical to promote understanding of concepts than to promote memorizing for the test, often with resulting forgetting of those skills. We should allow for multiple approaches and different solutions and strategies rather than look for only one answer. We should give students time to discover rather than penalize them for being slow. Assessment is an on-going part of every classroom and every activity. Teachers and students are all part of the assessment process, which will help students grow and learn as well as to appreciate the power of mathematics.

References

National Academy of Science. (1989). *Everybody counts.* Washington, DC: Author.

Wiggins, G. (1989). A true test: Toward more authentic and equitable assessment. *Phi Delta Kappan, 70*(9), 703-713.

Classroom Assessment That Informs Instruction: A CGI Teacher's Perspective

Mazie Jenkins
Madison, WI, Metropolitan School District

Assessment serves a variety of purposes: it can measure student achievement, it can monitor student progress, and it can be used to evaluate district mathematics programs. It can also be used to inform instructional decisions. There are a variety of assessment models that can be used to address each of these different purposes. This paper addresses assessment that measures student achievement to inform instructional decisions. This form of assessment is continuous and consistent with the principle that "student assessment [should] be integral to instruction" (National Council of Teachers of Mathematics, 1989, 1995). This type of assessment is embedded in Cognitively Guided Instruction (CGI). (The Appendix to this chapter contains a brief introduction to CGI.)

Because continuous assessment is embedded in the nature of CGI, CGI teachers are necessarily engaged in continuous assessment. Instructional activities provide an assessment opportunity for teachers as well as a learning opportunity for students. Teachers acquire detailed knowledge of each student's mathematical thinking and understanding, and use that knowledge as the basis for instructional decisions affecting individual students and the entire class.

Cognitively Guided Instruction is instruction guided by the teacher's knowledge of the cognition of individual students in the class. By requiring students to invent problem-solving strategies based on their familiarity with the problem context, and by requiring students to explain their solution strategies in detail, and by listening carefully to each student's explanation, teachers can make detailed descriptions of the thinking of each student in the class and can accurately predict what problems a student can solve, what size numbers are appropriate for that student, and what solution strategy the student will use for a specific problem. By understanding what makes some problems harder than others, teachers can then select harder or easier problems for students to work on.

When students in a CGI classroom solve problems, they reflect on their thinking orally, in writing, or both. These reflections assist the students in thinking about how they solved the problems and give teachers a window into the children's thinking. Assisted by the CGI framework of children's thinking, teachers can understand the students' strategies and analyze the child's thinking. Instead of saying only that a child can or cannot solve a particular type of problem, a CGI teacher can describe the strategy each child will use for any particular problem.

Continuous assessment provides the teacher with a way to analyze a student's thinking in order to inform subsequent instructional decisions. This process is a loop consisting of four main components: the teacher selects and poses a mathematical problem, the students engage in the mathematical activity, reflect on the mathematics of the activity and describe his or her thinking, the teacher listens to the student's description of his or her thinking, and the teacher assesses the student's thinking and provides a new task that appropriately challenges the student. This continuous assessment loop is characteristic of instruction in CGI classrooms.

To start the loop the teacher poses a task based on her knowledge of the students' thinking. On the first day of class the teacher has little information about the students and their thinking, so tasks early in the year may be selected on the basis of what she knows other students have been able to solve at this time of the year. With each new task, however, the teacher develops more extensive knowledge of the thinking the students, and revises her previous knowledge to incorporate more current information. While she cannot reassess each student each day, by judiciously selecting students to observe and listen to she can assess every student over a period of several days. By assessing daily, the teacher does not have to wait until the chapter test or unit test to know how the students are doing. By waiting until the end of the chapter, if the teacher finds that some students are not doing well, the time for instruction on that material has passed and it is now time for instruction on new material, material that frequently builds on the material that students have not yet learned. By assessing daily the teacher can make more informed judgments about how rapidly to move on to new topics.

In the second stage of the continuous assessment loop each student thinks about the problem situation and reflects on this or her prior experience with similar problems or similar situations. They solve the problem individually or by collaborating with other students.

In some CGI classrooms all students work on the same problem. In others, students work on the same problem but use different numbers. In still others both problems and number sizes may vary among students. What the classrooms all have in common is that the problems are rich enough that students at many different levels can work on and solve them in different ways. For example, students at any grade level, kindergarten through grade 12, could work on the problem: *Nineteen children are riding in a mini-bus. They can sit either two or three to a seat. There are seven seats. How many seats will have two children, and how many will have three?*

Many strategies are possible. Some are very elementary, but labor intensive. Some are very sophisticated, requiring more mental work and less physical work. Some possible strategies for this problem are depicted below.

Strategy 1: Many kindergarten children will count out 19 counters and group them into twos and threes until they have 7 groups and have used all 19 counters.
Comment: This strategy is called a direct modeling strategy. Children use counters to represent the objects in the problem, and directly model the action expressed in the problem. In a study of kindergarten students' problem solving ability (Carpenter, et al., 1993), 51% correctly solved this problem. In the same study 59% used a valid strategy. All but one student who used a valid strategy used a direct modeling strategy.

Strategy 2: Some children will draw 7 boxes to represent the seats and put 2 or 3 counters in each box, adjusting the numbers until each of the 7 boxes has either 2 or 3 counters in it.
Comment: This is also a direct modeling strategy, but it involves an explicit representation of the 7 seats.

Strategy 3: Some children will more systematically assign one counter to each of the 7 seats, then a second, and then a third until all counters have been used.

Comment: This is still a direct modeling strategy, but it involves a systematic assignment of counters to seats rather than a trial-and-error assignment as in the first two examples.

Strategy 4: Some children may be able to use addition or skip counting to apply Strategy 3 mentally. They could see that putting 1 child in each seat would use up 7 children, and a second in each seat would require 14. Since there are 5 children left, 5 seats would have 3 children, and 2, 7 - 5, would have 2 children.
Comment: This is not a direct modeling strategy since the seats and the students are not modeled as part of the solution.

Strategy 5: Another strategy involves using known facts. A child might know that 5 threes (5 x 3) is 15, so 5 seats with 3 on each seat is 15 students, leaving 4 students for the two other seats.

Strategy 6: A strategy taught in high school is to define two variables, x and y, to represent the number of two-person seats and three-person seats, and to create two equations that describe the relationships in the problem:

$$a) \ x + y = 7$$
$$b) \ 2x + 3y = 19.$$

By solving for x in equation a),

$$x = 7 - y,$$

using this value for x in equation b),

$$2(7 - y) + 3y = 19$$

solving for y,

$$y = 5$$

and using this value in equation a) to solve for x,

$$x + 5 = 7$$
$$x = 2$$

the problem is solved.

These are not the only ways to solve a rich problem such as this one. The variety of solution strategies, from very primitive direct modeling strategies to very elegant algebraic strategies, illustrates how students with a wide range of abilities can all work on the same problem successfully. It also illustrates how important it is for the teacher to be familiar with each student's strategy to understand his or her thinking, not just whether the child can solve the problem or not.

The teacher asks a child to explain his or her thinking in the third stage of the continuous assessment loop, either orally or in writing. The teacher's questions focus on the child's thinking and the reasons for the steps in the solution procedure rather than the answer or a recitation of the procedural steps. In this interaction the teacher might ask such questions as follows:

- Can you tell me your thinking?
- Tell me how you solved this problem.
- Have you ever solved a problem like this before? If so, how is this problem like that one? How is it different?
- Kendi, how is your strategy like Cassie's?
- Are these good strategies for solving this problem? Why or why not?
- How are your two strategies alike and how are they different (to a student who has solved a problem two different ways)?

Students' answers to such questions give teachers even more information about how they solved the problem. Teachers may find students' explanations hard to understand at first, since their strategies are frequently different than those used by the teacher. Careful listening is frequently required, along with repetition of the strategy by the student.

To complete the continuous assessment loop the teacher makes instructional decisions about the next problem to be posed, based on her most recent discoveries about the student's thinking and her accumulated knowledge from the entire year. The teacher must decide whether or not the last problem was appropriate. The strategy described by the child may indicate that the problem was not understood. Perhaps the numbers were too large or the context unfamiliar or too complex. Perhaps assistance in the form of scaffolding was needed. On the other hand, the problem may have been too easy. If the selection of number size is left to the child, the teacher might encourage him or her to increase the difficulty of the problem by choosing larger numbers. Otherwise, a conversation with the child may help the teacher select an appropriate problem.

One of the pivotal problems that assists teachers in making instructional decisions is the "join, change unknown" problem. This problem is characterized by an explicit joining action where the initial state and the final state are known, and the change that produces the final state is being sought.

> *Lonella had 9 books. Emmanuel gave her some more. Now she has 15 books. How many books did Emmanuel give Lonella?*

A more versatile version of this problem gives the students number size choices. In the problem below students can select the (a) numbers, the (b) numbers, the (c) numbers, or the (d) numbers.

> *Lonella had (a) 9 (b) 11 (c) 17 (d) 29 books. Emmanuel gave her some more. Now she has (a) 15 (b) 20 (c) 38 (d) 72 books. How many books did Emmanuel give Lonella?*

Here again, the multiple strategies possible for solving this problem give the teacher different information about the thinking of the students who use them. One student chose the (d) numbers, 29 and 72, and joined them to get 101. The teacher questioned her because she had solved a similar problem correctly on a previous day. The teacher asked the student to read the problem three times. On the second reading the student looked at the teacher and laughed. She then resolved the problem and recorded her thinking this way:

$$29 + 30 \longrightarrow 59 + 10 \longrightarrow 69 + 3 = 72$$
$$30 + 10 + 3 = 43.$$

The teacher then knew that the child could solve a join, change unknown problem, using ten as a unit. She also knew that the student was ready to try a more difficult problem such as a join, start unknown problem (a problem with a joining action in which the initial state is unknown but the change and the final state are known), a part-part-whole, part unknown problem (characterized by a static condition with no action in which one part and the whole are known and the other part is to be determined), or a compare problem (characterized by a static condition with no action in which the magnitude of two quantities is known and the difference between them is to be determined). She also suspected that this child was ready to work with three-digit numbers in problems.

A child who cannot solve this join, change unknown problem needs more problems that can be directly modeled such as join, result unknown problems, separate, result unknown problems, multiplication problems, measurement division problems, and partitive division problems. Over 70% of students in CGI kindergarten classrooms were able to solve multiplication problems, measurement division problems, and partitive division problems in the spring of the year (Carpenter et al., 1993).

There are four advantages to use of this continuous assessment model. First, the teacher gains insight into the child's thinking while posing appropriate questions during the listening component of the loop. Second, because the student explains his or her reasoning in considerable detail, the teacher has more information about the child's thinking on which to base her subsequent instructional decisions. Third, the teacher avoids making errors while analyzing student thinking by relying on immediate, direct evidence from the child through his or her verbal or written answers. Finally, because the teacher knows about each students' thinking through frequent interviews, she can describe the child's thinking and progress in detail to the child's parents.

This continuous assessment model has proven to be a powerful instructional tool, benefiting students and teachers alike. Because Cognitively Guided Instruction provides a framework for understanding children's thinking, it greatly facilitates a teacher's use of the continuous assessment model.

References

Carpenter, T. P., Ansell, E., Franke, M. L., Fennema, E., & Weisbeck, L. (1993). Models of problem solving: A study of kindergarten children's problem-solving processes. *Journal for Research in Mathematics Education, 24*, 428-441.

Chambers, D. L. (1994). Cognitively guided instruction. *Teaching Children Mathematics, 1*, 116.

National Council of Teachers of Mathematics.   (1989). *Curriculum and evaluation standards for school mathematics.* Reston, VA: Author.

National Council of Teachers of Mathematics.   (1995). *Assessment standards for school mathematics.* Reston, VA: Author.

Appendix: A Note on Cognitively Guided Instruction
(Chambers, 1994)

Cognitively Guided Instruction (CGI) is not a traditional primary school mathematics program. It does not prescribe the scope and sequence of the mathematics to be taught, provide instructional materials or activities for children, or suggest an optimal way to organize a class for instruction.

CGI started as a research program to investigate the impact on teachers and their students of research-based knowledge about children's mathematical thinking. Through an extensive series of professional development activities, teachers of grades K-3 learn about children's mathematical thinking and how that knowledge can help them understand their own students. In their classrooms, the teachers are then able to assess their students' thinking in great detail and make more informed instructional decisions that result in greater understanding and achievement by the students.

Even though CGI does not prescribe instruction, CGI classrooms do exhibit similarities. Children in CGI classrooms spend most of their time solving problems. Children are not shown how to solve the problems. Instead, each child solves them in any way that she or he can, sometimes in more than one way, and reports how the problem was solved to peers and to the teacher. The teacher and peers listen and question until they understand the problem solutions, and then the entire process is repeated. Using information from each child's reporting of problem solutions, teachers make decisions about what each child knows and how instruction should be structured to enable that child to learn.

In controlled studies, CGI teachers taught problem solving significantly more, and number facts significantly less, than control teachers. CGI teachers encouraged students to use a variety of problem-solving strategies, and they listened to the processes that their students used significantly more than control teachers. CGI teachers knew more about individual students' problem-solving processes, and they believed that instruction should build on students' existing knowledge more than did control teachers. Within the experimental group, the students of the teachers who were most influenced by the CGI professional-development program had the highest levels of achievement. At the end of the school year, experimental teachers' knowledge of their own students was significantly correlated with students' problem-solving performance.

Students in CGI classes exceeded students in control classes in number-fact knowledge, problem solving, reported understanding, and reported confidence in their problem-solving abilities. CGI teachers spent only about half as much time explicitly teaching number-fact skills as control teachers, yet CGI students actually recalled number facts at a higher level than control students. Studies in urban schools with large minority populations show similar effects.

With funding from the National Science Foundation from 1985 through 1996, CGI was directed by Elizabeth Fennema and Thomas Carpenter, Professors in the Department of Curriculum and Instruction, University of Wisconsin-Madison. The content domains included in CGI are addition/subtraction, multiplication/division, multi-digit numbers and base ten number concepts, and beginning fractions concepts.

Thoughts on Classroom Assessment

Jeane M. Joyner
North Carolina Department of Public Instruction

Translating theory into personal practice highlights the complexities of classroom assessment. It is so easy to talk about making stronger links between teaching and learning, yet so difficult to modify our classroom behaviors.

Part of the difficulty lies in the public's perceptions about the ways classrooms are supposed to operate. Teachers are to impart knowledge, usually the closer to the way it was learned by those discussing the topic the better. Another difficulty is the variation in the outcomes that are valued. "Basics" is the watchword of many communities, and often a consideration of what may be basic for the twenty-first century is not discussed.

A third difficulty lies in the roles we have ourselves experienced as teachers, students, and administrators. It is easy to take traditional roles. Teachers teach; students learn; teachers test; and the cycle begins again. Every teacher has numerous examples of this model from personal experience as a student. Besides, there is satisfaction and usually applause for playing the traditional roles well.

> *I've been teaching for a long time; my students like my classes and I feel that my children are learning. Students and parents are satisfied and often ask for students to be placed in my class. My students' test scores are good, and my own feelings of accomplishment attest to my success as a teacher.*

Into this comfortable scenario comes new information about how students learn and evidence about unreached potentials for all students. Research from cognitive sciences and neuroscience is challenging our beliefs about both teaching and learning. The profession is raising issues of appropriate content and assessments, the use of technology, concerns about equity, and the importance of self-assessment for students.

It seems clear that if classrooms are to become more powerful in supporting students' learning, many of us will need to modify our practices. This means a teacher-by-teacher movement that is informed by research, filled with interesting yet challenging content, grounded in self-reflection for both teachers and students, infused with appropriate models, and supported by school and system administrators. There is personal evidence, too, that suggests that what we often see in our classrooms are, to borrow an expression from Kathy Richardson, "illusions of learning." Working with the same students for several consecutive years is a humbling experience. From one year to the next much of what the students "learned" must have been put into short term memory, if indeed real learning took place at all. Many students seem to forget more than they remember.

Greater academic success and confidence in themselves as learners are goals that most teachers have for their students. However, the actions different teachers choose in order to achieve these goals vary greatly. Teachers who are comfortable questioning their own practices and who view themselves as life-long learners embrace change. For them, implementing new strategies related to teaching and learning are critical. For other teachers the information, ideas, and conversations about assessment stimulate interest but not necessarily action. Of course, new ideas and information are better received when they are presented in ways that do not cause the audience, in this case teachers, to erect walls of defensiveness. Unfortunately, there are also the inflexible educators for whom change will take place, usually poorly, only through mandates.

So what am I coming to believe about classroom assessment? There are abundant resources related to performance tasks, portfolios, alternative assessments in general. But I wonder if these strategies are being implemented without day to day influence over the manner in which teachers interact with students – with how well teachers engage with students in understanding clear goals and performance expectations, with how well teachers discern students' thinking and reasoning and make appropriate inferences and plans of action.

Participating in discussions about assessment and its purposes, whether with enthusiasm or hesitation, is a first step. Then for talk to become action there must be a willingness to question one's practices and reflect upon them in a manner that asks if there is a better way. A key to successfully transforming classroom practices seems to be to make our changing beliefs congruent with the ways in which we behave in the classroom.

What am I doing now that fits with what research says is powerful and matches my experiences and intuition? Are there strategies I could employ that will result in my students learning more? Will the changes be worth my effort? How can I find time to learn more about assessment, experiment with the ideas, meet the expectations of students, parents, and the administration, continue to teach every day, and still have a life? Where do I begin?

In the *Assessment Standards* (National Council of Teachers of Mathematics, 1995) the point is made that it is important to give thoughtful consideration to the different purposes of assessment and evaluation, since assessment for different purposes may take different formats and require different parameters. The four purposes discussed in the *Assessment Standards* include monitoring students' progress, planning for instruction, evaluating students' achievement, and evaluating programs.

While all four purposes of assessment impact the classroom, what needs greater attention and support is on-going monitoring of students' thinking and reasoning that informs teaching and learning. That is, assessment should drive teachers' instructional decisions and students' reflections about where they are in the process of attaining content goals.

Conversations about assessment inevitably move too quickly to assigning grades and evaluating achievement. Both teachers and students need to know where the students are in relation to goals that have been established. They also need to understand the standards for quality work, although at the end of each grading period teachers must make a judgment and mark each student's position in relation to the attainment of the goal. What informs instruction is not the marks but the many bits of information that led to the judgment. For example, teachers should be asking questions like the following:

- What meaning do I give to the actions of the students?
- What parts of a particular work are correct and what parts are incomplete or misunderstood?
- What concepts are still being developed and what ideas can serve as springboards for new explorations?
- What skills need reinforcing and what investigations will encourage synthesis through application?
- How do I promote meaningful self-assessment by the students?

Classroom assessment can become more powerful if we are guided by the root of the word assessment, "to sit beside." Our discussions need to focus on ways to help teachers develop insight into students' understandings, guidance in interpreting what is observed, and suggestions about acting on the inferences. We need to describe ways that the day-to-day interactions that inform teachers can also be used to give students feedback about their progress toward learning goals. We need to emphasize observation and conversation as important tools for teachers and develop strategies for helping teachers learn to use these tools.

These are steps in creating stronger links among content goals, instructional practices, and student achievement. Identifying what achievement looks like at the same time instructional plans are being made forces teachers and learners to be very definite about expectations. With clearly articulated learning targets, teachers may be better able to tailor the experiences they plan for their students.

> *Do not we recognize that teaching that moves from lesson to lesson without being strongly influenced by what the students are thinking is like shooting in the direction where targets are placed but being blindfolded as we begin shooting arrows? Pressures, perceived or real, that encourage teachers to march through materials, stopping occasionally to ask students to repeat what has been told them, are driving classroom practices in ineffective directions.*

Planning instruction and assessment together and modifying plans based on assessment information are fundamental to classroom assessment. But they are only part of a larger picture. How teachers go about gathering evidence of learning is another component of classroom assessment. Teachers need to understand advantages and disadvantages of selected response assessments, open-ended questions, performance assessments, observations, and all of the many formats assessment may take. How purpose and type of assessment relate, what portfolios and work samples provide, and what meaning can be given to the evidence that is collected are issues to be addressed.

> *Whoa! We still have not yet delved into helping students become more responsible for self-assessment or talked about rubrics or dealt with any number of other issues that always arise when assessment is discussed. But I am determined not to be overwhelmed. If reflection and self-evaluation are my first steps for creating the best possible scenario for my students, does not this say something about how other teachers learn and grow and change? Hold on to that idea. It may become part of a plan.*

Reference

National Council of Teachers of Mathematics. (1995). *Assessment Standards for School Mathematics*. Reston, VA: Author.

Using Information About and Results from Standardized Tests for Instructional Guidance in the Mathematics Classroom: Mixing Oil and Water?

Patricia Ann Kenney
University of Pittsburgh

My principal puts a great deal of emphasis on our school's performance on the state-mandated, standardized achievement test in mathematics. She is very concerned about how our students will do compared to other students in our school and, more generally, with students in others schools in the district and the state when the results are published in the newspaper. I'm trying to restructure how I teach mathematics based on the ideas in the NCTM Standards, but I'm worried that my students will not do well on the test because of its multiple-choice format and its emphasis on computation and procedures. It just doesn't match what I am doing and what my students have become accustomed to doing in mathematics. It's like mixing oil and water!!

A third-grade teacher offered the above comments during a recent workshop on assessment. His sentiments echoed those of each of the other twenty or so other teachers in that room. These teachers were very nervous about the standardized achievement test that was to be administered to all students in grades 3 through 11 during the next few weeks. The test has high-stakes consequences for students (e.g., placement into "skills improvement programs" for students performing below the 50th percentile), for schools (e.g., the expectation that a minimum of 50 percent of a school's students must perform at or above the 3rd quartile in total basic skills), and for districts (e.g., district's state-level accreditation to be based in part on students' performance on the standardized test). Moreover, school- and district-level results would be reported to the state legislature for review. Given these consequences, it was not surprising that every teacher at the workshop reported that the school principal had "strongly advised" allocating class time for students to practice the mathematics to be assessed on the test. Some teachers shared the following quotation distributed by their district testing supervisor to justify emphasis on the test:

> *[Other than in standardized testing], the norm of assessment in most other human activities, from manufacturing to athletics, is that one always teaches to the test. The basketball coach would never be told by the principal, "Your team may not practice, that would be cheating! Furthermore, you may not scout the other team or obtain films of the other team.... Just get your team ready to play by telling them, 'Don't think about the game at all. When the time comes, just do the best you can!'" No basketball coach would accept these conditions in a competitive basketball league. But we tell classroom teachers virtually the same thing all the time when it comes to practicing for the achievement test.*

It was also not surprising that the teachers wanted to vent about the situation and about how powerless they felt by having this test imposed on them and their students by the "powers-that-be." In fact, the teachers spent the first hour of the workshop expressing their anger, frustration, and concern -- and I, as the workshop leader, thought it wise to let this happen.

The concerns of these teachers that were just summarized -- in particular, their perceived degree of powerlessness with respect to the influence of externally-mandated standardized testing -- is not new. Using instructional time to prepare for and then to administer some kind of standardized test is a quite common occurrence in most mathematics classrooms in the United States. School districts or state departments of education frequently require students to take mathematics tests in order to use the test results for program evaluation and student placement, diagnosis, or summative evaluation. In many ways, the purpose of standardized tests -- to provide a means for comparing student performance on a predetermined but often somewhat restrictive set of objectives -- runs counter to the vision of the *Assessment Standards* (National Council of Teachers of Mathematics, 1995). That vision includes a shift away from assessing only students' knowledge of specific facts and isolated skills, comparing students' performance with that of other students, and basing inferences on single sources of evidence, and toward assessing students' mathematical power, comparing students' performance with established criteria, and basing inferences on multiple sources of evidence. The metaphor of "oil and water" mentioned in the opening quotation is very appropriate when one considers the mismatch between the purpose of standardized tests and this vision.

However, if one considers results from externally mandated standardized tests as only *one* of many sources of information about student performance in mathematics, then they can provide some information that teachers may find useful for the purposes of instructional guidance. For example, those tests can provide general within-student comparisons on different subjects (e.g., mathematics, reading, science) and across-student comparisons on the content assessed by the test (Silver & Kenney, 1995). At the very least, standardized test results may confirm what teachers already know about their students' understandings and misunderstandings about mathematics.

Because it is unlikely that standardized testing will disappear from the American educational scene in the near future and because evidence exists that teachers are influenced by the content and format of these tests (Madaus, West, Harmon, Lomax, & Viator, 1992), mathematics teachers and teacher educators may find it useful to examine ways that the frameworks and results from standardized tests can inform instructional practice, while at the same time realizing that the information derived from such tests will likely be very different from that available from classroom-based assessments such as observations, interviews, open-ended tasks, extended projects, and portfolios. The purpose of this paper is to foster an informed, in-depth look at standardized tests in order to understand better both their purpose and limitations and to examine how information from such tests can be used to inform instruction in the reform-oriented mathematics classroom. To accomplish this purpose, the remainder of the paper is divided into two sections: the first section deals briefly with the often dissonant relationship between standardized tests and classroom teaching practices, and the second section contains a description of an activity designed to involve teachers in thinking about how to use standardized test frameworks and results for instructional guidance, but in ways that support activities associated with the reform of school mathematics.

Standardized Testing and Teaching Practices

At some time during the school year (usually in the spring) every teacher must relinquish class time to administer a battery of standardized tests. Although there are many kinds of standardized tests, the focus of this paper is restricted to standardized tests that are nationally normed to allow for comparisons among students, that are exclusively multiple-choice in format, and that have high-stakes consequences for students. Silver (1992) addresses the pernicious effects attributed to high-stakes, externally mandated standardized tests in classroom practice. "The research suggests that teachers tend to narrow their instruction by giving a disproportionate amount of time and attention to teaching the specific content most heavily tested, rather than teaching concepts or overarching principles, or rather than teaching untested or less tested areas (e.g., geometry, data analysis) that are also expected to be part of the curriculum" (p. 492). In some circles, this phenomenon has been designated by the acronym WYTIWYG - "What you test is what you get." Silver further addresses the dilemma reform-minded teachers face, for whom teaching to a test not aligned with their curriculum and classroom practices is not acceptable, but who realize that performance on the test affects their students' educational future. Some teachers have found creative ways to balance the demands of the test while being true to their own practices. For example, some reports (e.g., Livingston, Castle, & Nations, 1989 [as reported in Silver, 1992]) suggest that teachers engage in a kind of "double entry" method that allows them to give sufficient attention to the goals of the standardized tests without sacrificing instructional attention to deeper conceptual understandings of the content. However, it is likely that such innovative methods have been adopted in only a small number of mathematics classrooms. (More complete information on assessment for instructional guidance in mathematics and on assessment and mathematics education reform can be found in Silver, 1992, and Silver & Kenney, 1995.)

Given the criticism directed toward externally mandated standardized tests, it seems unlikely that information about the test and the results from the test can be used to inform instruction in positive ways. Silver and Kenney (1995) acknowledge that externally mandated standardized tests offer teachers very limited information that can be used to guide instruction, but they also suggest that the results can provide classroom teachers with information on the content assessed on the test and student performance on that content. For example, as shown in Figure 1, the sample student report from the mathematics section of Stanford

Achievement Test, Ninth Edition (hereafter referred to as "Stanford 9") contains general information about student performance in particular mathematics content areas. The student in this example scored "above average" on measurement and patterns and relationships, but "below average" on geometry and spatial sense and computation in symbolic notation. One possible reaction to the report in Figure 1 is that it is far too general to be of immediate use to most teachers, and even raises more questions than the results can answer. For example, in the context of the Stanford 9, what does "geometry and spatial sense" mean? What do the test items look like? What specific concepts are assessed under the broad heading of geometry and spatial sense? Are these concepts in alignment with the vision of NCTM's *Curriculum Standards* (1989)? Asking these kinds of questions about standardized tests can enable teachers to understand better the purpose and limitations of such tests, thus "demystifying" the test and encouraging teachers to feel as if they had more control over the situation. The next section provides an example of an activity designed to acquaint teachers with the framework and objectives of a particular standardized test, to encourage them to operationalize the information provided on the reporting forms in concrete ways, and to think about how to use the information to inform classroom instruction.

	Below Average	Average	Above Average
Mathematics: Problem Solving		X	
Concepts/Whole Number		X	
Computation		X	
Number Sense and Numeration	X		
Geometry and Spatial Sense			X
Measurement		X	
Statistics and Probability		X	
Fraction and Decimal Concepts			X
Patterns and Relationships		X	
Problem-solving Strategies			
Mathematics: Procedures		X	
Number Facts		X	
Computation in Symbolic Notation	X		
Computation in Context		X	

Figure 1.   Sample Student Report - Mathematics Section of the Stanford Achievement Test Series, Ninth Edition

How Information from Standardized Tests Can Guide Instruction

The opening paragraphs of this paper outlined the concerns of a group of teachers attending a workshop on how a standardized test -- in this case, the Stanford 9 -- could provide information to instruction. Based on the premise that standardized mathematics tests can provide teachers with some information based on performance on the content areas assessed on the test, an activity was designed to make the information provided by the Stanford 9 student reports more concrete, thus encouraging teachers to think about what a test question written to address a Stanford 9 objective might look like, how the mathematical concept was assessed, and what misconceptions the question could potentially uncover. Working with a set of items that matched the Stanford 9 objectives could serve to demystify the test and to encourage teachers to develop creative ways to assess how their students were performing on the test's objectives -- especially those objectives that matched important content within their own classrooms -- and then based on students' understandings or misconceptions, to think about how this information might impact their instructional practices.

Because actual Stanford 9 items were not available in a timely manner, a source of items that was similar to those on the test had to be identified. Fortunately, a potential pool of items did exist -- the released questions from the 1992 National Assessment of Educational Progress (NAEP) in mathematics. These items were classified according to content area categories and subcatgories in the NAEP objectives (NAEP, 1988). A preliminary examination of the Stanford 9 objectives and the NAEP objectives suggested that in many cases the objectives statements were nearly identical. For example, for grade 4 both sets of objectives had statements about rounding whole numbers and reading/interpreting bar graphs; and for grades 8 and 12 both sets had statements about evaluating linear expressions and calculating areas of plane figures. Given the similarities between objectives from the Stanford 9 and NAEP, it was not unreasonable to suppose that a multiple-choice item developed to fit a content category in the NAEP framework might be similar to a Stanford-9 item in that same category. (It would be useful to test this conjecture more thoroughly by systematically comparing the Stanford 9 framework and the NAEP framework and by comparing sample items from NAEP and from a set of released questions from the Stanford 9.)

The activity with teachers was built around released questions from the NAEP assessment at grades 4, 8, and 12 that matched at least one objective on the Stanford 9. (A description of the activity appears in the appendix to this paper.) First, the teachers working in pairs or groups answered the question. Because not all of them were mathematics teachers, they were asked to pair-up with another teacher currently teaching mathematics. Next, they matched the question to the Stanford 9 objectives statements. The purpose of this part of the activity was to encourage teachers to make the link between the question and the test content. (It was at this point in the workshop that I received some anecdotal evidence that the NAEP items were indeed similar to those on the Stanford 9. While working on the activity, a mathematics specialist who had seen the Stanford 9 practice items at grade 4 commented that the grade-4 NAEP items used in the activity were very typical of the kind of items on the Stanford test a grade 4.) The last part of the activity -- examining distracters and thinking about the misconceptions embedded in them -- was considered to be the most important part of the activity itself. In fact, this part stimulated the most lively discussions among the teachers. The discussions were very animated, and all teachers (even those who did not teach mathematics) were quick to identify misconceptions that would have students choosing 18,600 as the answer to the fourth-grade question on rounding or that would have students choosing 512 instead of 1,024 as the answer to the eighth-grade problem on volume. (The reading teachers were vocal about this problem, saying that despite the underlining of the key word *half*, students would still skip over it.) During the course of the activity, the teachers began identifying some distracters as "good wrong answers" (that is, incorrect answers based on common misconceptions) and talked about classroom-based activities that would address the misconceptions associated with these distracters.

The teachers also talked about ways to use these sample questions in their classrooms. One teacher suggested giving the questions to students and have them react to parts of the activity from the workshop; in particular, identifying the mathematics concept assessed by the question and talking about the distracters and the misconceptions they illustrate. Another teacher shared an activity similar to the one from the workshop. In her high-school algebra class, multiple-choice questions from a sample Stanford-9 item set became the basis for a series of warm-up activities. As the students entered the classroom, they took a question from the envelop near the door, worked on solving the question and showing their work, and then as an on-going journal assignment, they wrote about the question, addressing the mathematics assessed, their solution strategies (other than guessing!), and the misconceptions associated with the distracters. At times the students had opportunities to talk about these questions and to share their journal entries. This teacher reported that the students liked this activity, and she felt that it fulfilled the directive to "practice for the test" but in a way that did not compromise her own reform-minded classroom practices.

Preliminary feedback on this activity suggests that the teachers attending the workshop thought it worthwhile, and that -- at least in the opinion of the mathematics teachers -- it had the potential to be the operational link between the information on the Stanford 9 reports and how to use that information to inform instruction. As a side note, the reading teachers attending the workshop requested a similar activity using the released NAEP reading items!

Conclusions

Although some might wish otherwise, externally mandated tests on which the majority of the questions (if not all questions) are in multiple-choice format are not likely to disappear from the educational scene in the near future. As teachers continue to receive information about their students from externally mandated tests, it becomes even more important for them to understand the test content and the reporting methods, and to devise ways to think about how to use the information to inform instruction. As one teacher in the workshop noted, "The activity we just finished confirms the need for me to be a critical consumer of information from standardized tests. For my students' sake, I don't want to completely ignore these tests and the reports -- especially if they can help me help my students."

As expressed by the teacher quoted at the beginning of this paper, "oil and water" may be an appropriate metaphor for the relationship between externally mandated standardized tests and the vision for assessment reform in the mathematics classroom. However, in the case of oil and water, adding soap to the solution makes the solution more homogeneous. Perhaps activities like the one described in this paper and other activities designed to demystify standardized tests can become the "soap" that assists teachers to understand better the purpose and limitations of these measures, and to make instructional decisions based on results from tests as one of many sources of information about their students' mathematical power.

Acknowledgment

The idea for this paper came during a workshop designed for teachers in the Benedum Collaborative Professional Development Schools, held on March 21, 1997, at West Virginia University. I am grateful to Jaci Webb-Dempsey, Sarah Steel and their colleagues at the Benedum Center for Educational Reform at West Virginia University, and the teachers who attended the workshop and who share their ideas and opinions so willingly. Preparation of this paper was supported in part by a grant from the National Science Foundation to the National Council of Teachers of Mathematics (Grant No. RED-9453189). In particular, the NAEP materials used in this paper were obtained through my affiliation with that project. Any opinions expressed herein are those of the author and do not necessarily reflect those of the National Science Foundation or National Council of Teachers of Mathematics.

References

Livingston, C., Castle, S., & Nations, J. (1989). Testing and curriculum reform: One school's experience. *Educational Leadership*, *46*(7), 23-25.

Madaus, G. F., West, M. M., Harmon, M. C., Lomax, R. G., & Viator, K. T. (1992). *The influence of testing on teaching math and science in grades 4-12*. Chestnut Hill, MA: Center for Testing, Evaluation and Educational Policy, Boston College.

National Assessment of Educational Progress. (1988). *Mathematics objectives: 1990 assessment*. Princeton, NJ: Educational Testing Service, National Assessment of Educational Progress.

National Council of Teachers of Mathematics. (1989). *Curriculum and evaluation standards for school mathematics*. Reston, VA: Author.

National Council of Teachers of Mathematics. (1995). *Assessment standards for school mathematics*. Reston, VA: Author.

Silver, E. A. (1992). Assessment and mathematics education reform in the United States. *International Journal of Educational Research*, *17*, 489-502.

Silver, E. A., & Kenney, P. A. (1995). Sources of assessment information for instructional guidance in mathematics. In T. A. Romberg (Ed.), *Reform in school mathematics and authentic assessment* (pp. 38-86). Albany, NY: State University of New York Press.

APPENDIX: Description of the Workshop Activity

Purpose: The intent of this activity was to provide teachers with experience in looking for mathematics questions that match the objectives of the Stanford 9. Once teachers found a question that matches at least one objective, then they could give it to their students and use information on performance to guide instruction.

There were two pre-selected sets of questions available in the workshop: one set geared toward the objectives for the elementary grades (specifically, grade 4) and the other, toward the objectives for middle/high school (specifically, grades 8 and 12). Teachers selected a grade level and then worked with the materials appropriate for that level.

Practice Activity: Working in small groups, the teachers examined the mathematics questions shown below and completed the following questions:

1. According to the Stanford 9 Objectives Statements, is the question Mathematics Problem Solving or Mathematics Procedures?
2. Which content objective best matches the question?
3. Examine at least two distracters (incorrect answers). For each distracter:
    a. What misconception might lead students to choose that answer?
    b. What classroom activities could you create to correct that misconception?

Activity Using Other Questions: Working in pairs, the teachers repeated the above activity using another set containing 6 to 8 questions.

Discussion: The teachers shared their work with the group, focusing on how they would remedy the misconceptions that they identified from the distracter analysis (i.e., the analysis of incorrect choices).

Grade 4 Practice Question

Solve the problem below and choose the best answer.

What is 18,565 rounded to the nearest thousand?

A. 18,000   B. 18,600   C. 19,000   D. 20,000

Grades 8/12 Practice Question

Solve the problem below and choose the best answer.

It takes 64 identical cubes to *half* fill a rectangular box. If each cube has a volume of 8 cubic centimeters, what is the volume of the box in cubic centimeters?

A. 1,024   B. 512   C. 128   D. 16   E. 8

Thoughts on Classroom Assessment

Mary Montgomery Lindquist
Columbus State University

The process of assessment is an amalgamation of the processes of measurement and statistical problem solving. The process of measurement requires us to understand the attribute or domain we are measuring. The process of statistical problem solving requires us to make sense of data, often gathered from many sources. Assessment is a complicated process for which we need methods and techniques and sound knowledge of the domains we are measuring.

The domain under consideration for this paper includes mathematics as well as the dispositions. Defining this domain is not a simple task. The National Council of Teachers of Mathematics (NCTM) *Standards* (1989, 1991) give a broad vision of the mathematics and dispositions; each state has refined that vision; and local districts and schools often have their own guidelines. Definition is necessary but not sufficient. At the classroom level, it must be clear what we expect and what we are measuring at even a more specific yet deeper level than any of the levels of standards or benchmarks. In measuring a physical attribute such as area, if we do not understand the attribute then we may arrive at a number that does not describe the area. In measuring a student's understanding of mathematics, we need a deep understanding of that mathematics. Similarly, in measuring dispositions, we need to understand those dispositions and their influence on learning mathematics.

After we identify that we are measuring the attribute of area, we then select a unit to which we compare the region we are measuring. In classroom assessment, after we specify the mathematics or disposition we want to measure, then we must select a standard to which it is compared. In the past, we have often made this comparison to other students: Did Jamie do better or worse than Pat? In looking at standards-based assessment, we make this comparison to levels of performance. As in measuring the area of a region, we often need instruments or methods to make that comparison. From this we arrive at a description which may or may not be a number. For example, the area is smaller than the space we want to fill, or the area is about 15 square units. The student has not reached the expected level of understanding, or the student scored 93 on a test.

It is clear that we know much more about measuring physical attributes than we know about measuring mathematical skills and understandings. We have no absolute standard as a square centimeter; we add unlike units to give a total score; and we average dissimilar scores to assign a number often with unwarranted precision. Furthermore, if we find a number but want to know more about the shape of the plane region, a number tells us little. Similarly, a final grade tells us little about what the student actually knows and can do. Thus, we turn to the lessons from making decisions in a more statistical way.

When we solve a problem statistically, we must be able to state the problem we are solving. In the case of assessment, the problem includes the purpose of assessment. The *Assessment Standards* (NCTM, 1995) provides a thorough discussion of purposes of assessment. We also need to know what assumptions we are making. We need to know if the measurement is representative of the student's performance. We need to know how to analyze often a diverse set of measures. To make instructional decisions, we may solve the problem in a less formal way than we do for other purposes. For example, if the purpose is to assign a grade or to compare a student to a standard, we may need more systematic and defensible evidence. We gather information in a measurement sense, and we make decisions in a statistical sense.

The two criteria that I attempted to keep in mind when addressing the questions posed for the conference were the *complexity* of the process and domain and the *realities* of classrooms and teachers' lives. My first reaction as I read the questions was that too much is written already about assessment. Yet, when I attempted to wrestle with the questions, I found I knew few references that would reflect these two criteria. The remainder of this paper addresses the first question posed for the conference with some sketchy thoughts regarding the other questions.

What kind of information do teachers need to provide to students about the learning targets? (How do teachers establish expectations and goals? What is the role of the student?) Is this information dependent on grade level?

Before attempting to discuss these questions, let me convey the sense of a discussion of students in a master's level secondary mathematics class. The first question did not raise the issues that I had expected. It was almost nonsense to them. They saw little to answer; they had goals and expectations, and they told these to their students. The class, which consists of seven teachers who are presently teaching mainly in rural areas, was comfortable with the fact that the objectives (their goals) were in their texts. Most stated emphatically that they were required to write the objectives on the board each day. On further questioning, it was evident that the learning targets were topics; for example, adding integers or factoring trinomials. A previous discussion that night in class had raised the issue about the vagueness of the local benchmarks; one teacher felt she did not know what to emphasis or even what many of them meant.

There was no discussion that the information depended on the grade level, but as I listened to the small groups discuss, this part was dismissed as an irrelevant question. In fairness to the class, the teachers were probably only thinking about a high school setting, and we did not pursue it further. The question about the role of the student triggered a discussion of classroom management and students setting expectations for the environment of the class. This, in many ways, was a richer discussion than the one regarding learning targets. To this class, setting expectations was much more about management and overall expectations (homework, grading, and the like) than about learning content. After reflecting on this discussion, it became more clear to me how the teachers are realistically trying to make sense of what they are doing in their busy lives.

We need to have teachers develop a usable set of learning targets and a picture of how they all can fit together. We are lost in the forest without a map and are seeing only a few of the trees. Some of these trees are beautiful, some are familiar, some are approachable, and some are usable; this keeps us going on our journey. In other words, the reliance on small, daily objectives mainly focused on procedures and skills is comforting to many teachers. Without a broader view of mathematics and a knowledge of mathematical learning, this focus will remain the norm.

Returning to the first groups of questions. Teachers need to have clear goals regarding the content before they can convey to students those goals. Obvious? Certainly, but this is not the norm. Without this picture, mathematics is often conveyed as a bunch of little things, some of which are learned each day. Perhaps, providing to the students an unfinished picture of the big ideas that they would complete by filling in the web would be more powerful than the practice of having a notebook in which the objectives for the day, the definitions, and the rules are copied from the text.

Although one can verbalize expectations, they are probably conveyed more subtly through actions of the teacher and samples of acceptable and exceptional works. Content expectations are facilitated through the expectations for the classroom environment, and thus this part of the conversation in my class became much more central. I wonder if the teacher who had his students set expectations (or rules) for the class and who was successful in having students take ownership and pride in these expectations, also had them set expectations for learning. If not, it would be an interesting experiment for him. Would they look at expectations that all the students should learn, should participate, and should help each other? Would they include persistence in solving a problem, pride in the presentation of a project, and pleasure in being challenged? Would they insist that mathematics should make sense to them?

Returning to the process of measurement. The setting of goals and expectations is analogous to understanding the domain to be measured or to be assessed. Without this understanding, the measuring is meaningless or, at best, is at the level of a simple comparison.

Returning to the process of statistical problem-solving. The next four questions posed for this conference focus more on this aspect of assessment. What is evidence? How do we gather it efficiently and consistently? How can we organize and interpret the evidence? And, how can we report that evidence in a meaningful manner.

What kinds of information about students do teachers need to gather in their daily interactions with students? What constitutes quality evidence? Is this information dependent on grade level?

The first question in this set needs to be answered but it needs to be put in the larger picture. Gathering information about students daily is like watching a child grow daily. You must stand back and wait; you do not need to measure them each day. The information gathered daily may be to give the sense of the group of students as a whole with targeted information about individuals - different individuals on different days. It is the composite of the evidence that we need to make decisions.

In discussing quality evidence, one must consider reliability (consistency) and validity (worthwhile mathematical tasks that fit into a whole picture). Once more, the decision that is to be made should influence the time and effort in gathering this evidence.

There is a struggle with the amount of evidence that someone who works with 150 students can make and use wisely. The feature of our schooling in which we gradually wean the students from a single teacher may be a good quality. Older students need to take more responsibility in monitoring their progress, knowing their own strengths, and responding accordingly.

What techniques seem most useful in helping teachers gather that information and what skills do teachers need in order to implement these strategies? Are useful techniques dependent on grade level?

We all need training in listening and ways to filter the vast amounts of information that we are receiving. At the lower grade levels, the vast amount of information comes from knowing many, many things about each student. At the upper grade levels, the information comes from the many students. In either case, the skill needed is knowing which data are relevant to helping students learn mathematics.

How can teachers learn to organize and interpret that evidence appropriately in order to make sound judgments about student learning and achievement?

The first question that calls for an answer is how teachers can organize and interpret that evidence, then we can turn to the question of how they can learn to do it. We need to observe those that are masters of making sound judgments. As we think about this aspect, we should think about the role of outside assessment, on-demand, that will give credence to our classroom sound judgments.

What constitutes quality feedback and in what ways can teachers provide useful feedback to students and their families?

These are the questions that teachers really want answered. Keeping in mind the realities of the lives of teachers, we need examples of ways this can be done in a meaningful way. In some ways, if the other questions posed have reasonable solutions, I think this one will follow. My recommendation would be to work on the others first and then return to this.

What is known about helping teachers learn to make instructional decisions based on that evidence?

I am not very confident that we know much about how teachers assess or use assessment other than broad generalizations or specific anecdotal records. This is an area that is fertile for research. It does seem to me that we can learn from the research from elementary programs such as Cognitively Guided Instruction (e.g., Carpenter & Fennema, 1992) and others. Important characteristics to be considered are the knowledge that teachers have of the domain, both the mathematical and learning domains; the high but reasonable expectations they set for students; and the lack of distinguishing assessment from teaching for the purpose of making instructional decisions.

The questions posed for this conference were deep and thoughtful. Even after attending and reflecting on the conference and my other experiences, I feel inadequate to answer them except on a superficial level. Research and examples are in order. Let us hope that we continue to discuss and to learn as we try to bring our knowledge of measurement and statistics to the process of assessing students' understanding of mathematics.

References

Carpenter, T. P., & Fennema, E. (1992). Cognitively guided instruction: Building on the knowledge of students and teachers. *International Journal of Research in Education, 17*(5), 457-70.

National Council of Teachers of Mathematics. (1989). *Curriculum and evaluation standards for school mathematics.* Reston, VA: Author.

National Council of Teachers of Mathematics. (1991). *Professional standards for teaching mathematics.* Reston, VA: Author.

National Council of Teachers of Mathematics. (1995). *Assessment standards for school mathematics.* Reston, VA: Author.

Classroom Assessment: View from a First-Grade Classroom

Carol Wickham Midgett
Southport, NC, Elementary School

Assessment is the vehicle for establishing mathematical literacy. Learning occurs when prior knowledge and experience are affirmed and/or adjusted by new knowledge and experience. Assessment is the means by which we determine the impact of this transformation. When assessment occurs in the midst of, as well as at the end of, mathematical learning events, students use misapplications to nurture growth in their understanding of content and process rather than being "punished" for not knowing and not being able to do.

The value assigned to assessment determines its purpose and role. Therefore, we must understand the relationship between assessment and the learning of mathematics. If we truly want students to be mathematically literate, we will structure assessment so teachers serve as researchers collecting data from observations, interviews, student products, and on demand performances. Teachers analyze these data to understand a learner's conceptions and misconceptions in order to make decisions about the mathematics curriculum, instruction and further assessment practices.

If we value this view of assessment and trust teachers to make "professional" decisions, classroom structures and roles will change. Members of the learning community will share the responsibility for and be accountable for the development of mathematical literacy. Teachers will no longer only deliver information. They will establish learning targets for mathematics, provide meaningful experiences in which the learning goals can be met and continuously monitor progress to adjust instructional strategies and resources.

When teachers formulate mathematical learning targets, they are accountable for developing mathematical literacy. In preparing short-term and long-term learning goals, teachers establish reasonable expectations based upon identified student needs and the curriculum, be it generated in the classroom or prescribed. They demonstrate a genuine understanding of knowledge, procedure, and instructional strategies through carefully planned mathematical learning experiences.

Teachers who focus on helping children become mathematically literate reflexively collect data to provide feedback and make judgments. Coming to know students' mathematical understanding and skill requires teachers to listen to and observe them in learning contexts. As teachers monitor student performance, they listen for children's mathematical thinking and reasoning. Teachers need to know the students' level of understanding of mathematics content and process in order to identify appropriate or inappropriate conceptual and procedural knowledge. Listening to student responses exposes prior knowledge and connections the learner is making between "old" and "new" information. Quality evidence is gained from interviews, observations, collections of student work, student reflections, and on-demand performances. Collecting and recording data from multiple sources provides a picture of what the student knows, is learning and needs to know.

In order to collect meaningful data, teachers need to engage students in mathematical investigations involving important concepts in real contexts. Teachers need to ask key open-ended questions that encourage students to think about their own learning.

Teachers also need to record data in ways that are useful in making formative and summative decisions. Each teacher has to develop a system of record keeping that is compatible with her/his personal management style. Suggestions from others are essential, but individual systems are rarely successfully duplicated. A major component of effective record keeping is for teachers to view assessment as a vital component of the learning process. This enables them to recognize that time spent assessing is a natural and essential part of teaching and learning. Teachers must know that keeping records for assessment of mathematical literacy is instructional for students, teachers, parents, and administrators. Teachers who engage in on-going assessment are able to make informed instructional decisions when they reflect upon assessment data, learning targets, and program expectations in relationship to students' demonstrated level of proficiency. Consistently engaging in this recursive process maximizes learning and leads to mathematical literacy.

Teachers who make assessment an integral part of their instructional planning for mathematics classes are more effective facilitators of learning. When teachers establish targets before engaging students in mathematical investigations, they know what to expect students to know and be able to do. Therefore, learning tasks are selected because they have a specific focus. They are structured to meet the needs of students, instruction, and programs. These mathematics learning events create a higher level of mathematical literacy. Under these conditions assessment is planned and data are purposefully collected. This information yields specific, relevant information related to the learning targets. This feedforward approach to learning yields many benefits for students.

The role of students changes when assessment is a cohort of instruction. Students assume responsibility for their own learning when they are expected to invest in it. Having clear, complete learning goals develops student confidence and competence in mathematics. Learners have the "big picture" when they know what is expected of them and know how it will be evaluated. This knowledge provides a framework for planning and defines available support. It establishes perimeters and suggests processes. When students know what is expected of them, they are able to clarify unknowns prior to engaging in a task. This preparation allows students to demonstrate what they know and are able to do. It makes them accountable and responsible for their learning.

Students develop confidence and competence as they engage in reflection. As students have opportunities to use and develop scoring guides for measuring achievement in mathematical investigations, they acquire independence in making judgments about their own work. Constant monitoring and immediate feedback support each student's ability to measure personal success. When students report their own progress, they demonstrate a higher level of accountability.

Student-led conferences are celebrations of learning. They are collaborative events in which students assume responsibility for presenting evidence of learning to their parents. As students participate in this process, they develop an appropriate understanding of learning expectations. They develop and use criteria for monitoring and measuring progress as they select products and events to share with their parents during conferences. Students engage in aligning expectations and evidence of progress. They formulate expository and defensive statements about their achievements and goals. Thus, students develop life-skills as they prepare for and lead conferences. They are responsible for and have control of their own learning.

Students who take an active role in their own assessment are truly equipped to be life-long learners. They recognize standards, establish criteria for measuring their own achievement, set personal learning goals, report growth toward mathematical literacy, and assume responsibility for their own learning. They learn how to determine when they have learned.

The role of parents changes when assessment is an integral part of the classroom learning process. Parents no longer only react to a single test or measure of a child's mathematical proficiency. Parents have multiple evidences of what a child can do whereas before the emphasis was on what the child was not able to do. Parents understand learning targets, recognize standards of proficient performance, observe their child's progress through collections of evidence, and accept responsibility for facilitating their child's achievement. They are prepared to be accountable members of the learning community.

Student-led conferences are exciting ways of providing quality feedback to parents and nurturing the accountability of students, teachers, and parents. All participants discuss learning expectations and achievement. Conferences led by students focus on continuous mathematical growth over time. Performance is compared with previous demonstrations rather than comparisons among students. For this reason, student-led conferences provide equity for each child.

When we view assessment as the vehicle for establishing mathematical literacy, new issues challenge us. Professional development should focus on helping teachers do the following:

- Become mathematically literate
- Incorporate assessment in the classroom learning process
- Learn how to articulate learning targets
- Develop a disposition for facilitating learning
- Master coaching skills
- Engage in action research
- Generate meaningful mathematical tasks
- Expect students to share progress with parents

Challenges for students are:

- Assuming responsibility for their own learning
- Recognizing learning targets
- Reflecting on their own mathematical learning
- Reporting academic progress
- Developing scoring criteria

Challenges for parents are:

- Knowing learning targets
- Learning content and procedural proficiency
- Attending conferences
- Meeting learning goals they set
- Supporting student learning

Challenges for administrators are:

- Acknowledging teachers as learning leaders whose inferences facilitate learning
- Publishing content expectations and characteristics of proficient learners
- Accepting multiple sources of evidence as valid measures of learning
- Supporting shifts in curriculum development and instructional design
- Providing time for teachers to collect data and reflect upon its implication
- Aligning the evaluation of teachers with this view of assessment of children
- Expecting parents to be involved in the learning community
- Facilitating appropriate professional development

If all of this happens, we will have students who value mathematics; think, reason, and communicate mathematically; make connections between mathematics and real-world situations; and are confident and competent problem solvers. Quality learning will occur and students will become mathematically literate when members of the learning community know the mathematics learning targets, participate in meaningful learning, engage in the development and implementation of measurement criteria, assume appropriate responsibility for learning, and set appropriate learning goals. Classroom assessment is indeed the key to creating mathematical literacy.

What Teachers Want To Know About Classroom Assessment

Vicki Moss
Wake County, NC, and Randolph County, NC, Schools

> *"Assessment and evaluation are interrelated but teachers don't understand the differences."*
> Instructional Resource Teacher, Wake County Schools

This year many discussions with our lead teachers have focused on the role of the teacher and learner as it relates to assessment practices. The comment quoted above, made all too often, cannot be ignored. Teachers are struggling with classroom assessment, not because they see no merit in the process, but because they lack the understanding of how to make it work to their advantage. Teachers believe that on-going classroom assessment can have a profound impact on how and what they teach. Many think on-going assessment enables them to observe and understand a student's thinking much more effectively than any other form of assessment but, they go on to say: *It just takes too much time.*

Several questions continually come to mind as I attempt to provide teachers with the support needed for implementation of successful classroom assessment practices. The first question deals with expectations: *What do I want my students to know and be able to use in various situations?* For teachers in North Carolina much of this question is answered in the *Standard Course of Study*, which is the required curriculum to be taught in all local education agencies. In addition, the North Carolina End-of-Grade Testing Program, a summative assessment instrument, provides the level of a student's performance on items written to this curriculum in the areas of reading and mathematics. Beyond the goals and objectives that make up the *Standard Course of Study*, teachers need to be able to see where students' learning will ultimately take them and get beyond isolated lists of skills and concepts. This requires that teachers not only be knowledgeable of curriculum, instructional methodology, and child development, but also be acutely aware of the diverse backgrounds and learning styles of their students.

Communicating these expectations to students is an essential part of the process. Enabling teachers to find ways to better communicate expectations and give quality feedback begins with the youngest learners. Students need to be given opportunities to go through the process of evaluating their own work. This may come about through teacher discussion with the class or individual but should include samples of work that meet the expected outcomes. When teachers are asked to think about assignments they were given in undergraduate or graduate school, they often indicate that the first question that comes to mind is, "What does the work have to look like to get an A?" Students, regardless of age, need to be given an opportunity to know what must be done to be considered proficient on a task. Classroom assessment is about all of the players' knowing the expectations.

The second question, and the one that seems most troublesome, deals with performance: *How will I know it when I see it?* Teachers must know what they are looking for. More than ever, teachers must become daily researchers. Students may exhibit their understanding through written assessments, interviews with teachers and peers, questions they ask, and a variety of products and performances. Just as students need to know what is expected, teachers must take a close look at what information students give them. What do they expect the top paper or response to be? What key elements will it contain? Rubrics can be used in conjunction with work samples to help teachers identify the key learning they are assessing on any task. Much of the information needed to make a thorough assessment of a student's understanding is readily available to teachers but often difficult for them to see. Many times their fear seems to be that someone else may think their expectations are not high enough or the wrong information was assessed.

In seeking ways to assist teachers in knowing how individual students are performing relative to curricular objectives in mathematics, our district recently implemented the use of a mathematics matrix similar to the sample shown in Figure 1. Teachers are to mark students' progress throughout the year and collect samples of work that document students' learning. What the documentation should be has become a major concern. The questions are most often twofold: How many times do I need to see something before I can mark a student at the proficient level (mastery)? What does the documentation need to look like?

Multiple methods of assessment offer a rich and broad look at a student's understanding. It is still acceptable to use chapter tests as a means of assessing understanding, since they provide very different information from an in-depth investigation into a topic that requires applying previous knowledge in another situation. Observing a student interact with a partner, in a small group of peers or in a whole class setting may also give the teacher better insight into what the student understands and is able to use in some meaningful manner. It is important to recognize that some students who are high performers in one setting may perform much lower in another. Recognizing a learner's strengths and capitalizing on them best enables instruction and assessment practices to reflect accurately what the learner knows.

Many factors determine the "what it looks like" and require the teacher not only to listen to and observe each student but also to assess performance for the individual learner based on where they started and how far they have come. Documentation, therefore, is time consuming, but necessary. Being selective is the secret to making record keeping manageable. Teachers who know what the instructional goals are and plan for meaningful assessment, integrated into daily learning, can prevent documentation from becoming an "add-on" or unreasonable task.

The last question that continues to be asked, after I know what I want to see and I know what it looks like, deals with next steps: *What do I do about it now?* Being able to observe and interact with students as they engage in meaningful learning activities provides the best view of what a student actually understands and is capable of doing. Often students are limited in the scope of activities that are provided in a classroom setting. Many times learning activities address one learning style, therefore preventing some students from being successful at the task from the beginning. Tasks that are too narrowly focused provide minimal information into a student's understanding. In addition, if the entire class is taught something in the same way and is expected to respond to the task in the same manner, it will be impossible to recognize what individual students know.

	Performance Indicators	Numeration	Geometry
Level IV	• consistent performance beyond grade level • works independently • understands advanced concepts • applies strategies creatively • analyzes and synthesizes • shows confidence and initiative • justifies and elaborates responses • makes critical judgments • makes applications and extensions beyond grade level • applies Level III competencies in more challenging situations	• reads, writes and counts beyond 100; reads number words beyond 10 • recognizes sets to 5 without counting • identifies original positions beyond tenth • compares, sequences numerals beyond 100 • skip counts by 2s, 3s, 5s, 10s and relates to repeated addition • can predict patterns beyond 10s and 1s	• creates models of plane and solid figures • identifies, makes figures with line symmetry • matches congruent figures • replicates 3-dimensional designs using models • groups by attributes geometric figures
Level III (Proficiency)	• **exhibits consistent performance** • **shows conceptual understanding** • **applies strategies in most situations** • **responds with appropriate answer or procedure** • **completes tasks accurately** • **needs minimal assistance** • **takes appropriate risks** • **makes applications** • **exhibits fluency** • **shows some flexibility in thinking** • **works with confidence** • **recognizes cause and effect relationships** • **applies models, and explains concepts**	• **uses counting strategies; 1-to-1 correspondence, counting on, tallying, grouping** • **makes, compares, orders sets and numerals** • **identifies ordinal positions** • **conserves numbers** • **reads, writes, represents numbers in a variety of ways; reads number words 0 to 10** • **recognizes one more, less, before, after, between** • **rote counts by 1s, 10s, 5s, 2s** • **makes reasonable estimates of "how many"** • **groups objects into tens and ones; record** • **recognizes models; builds 2-digit numbers; writes numerals**	• **identifies open and closed figures** • **identifies, describes, models plane figures** • **describes likenesses and differences (i.e. circles, squares, rectangles, triangles, hexagons, trapezoids)** • **identifies, describes solids (i.e. cubes, cylinders, spheres, rectangular prisms)** • **recognizes examples of plane, solid figures in the environment** • **uses comparative, directional, positional words**

Figure 1. First Grade Observation Matrix for On-Going Assessment and End-of-Year Evaluation (Figure continued on next page)

	Performance Indicators	Numeration	Geometry
Level II	• exhibits inconsistent performance and misunderstandings at times • shows some evidence of conceptual understanding • has difficulty applying strategies in unfamiliar situations • responds with appropriate answer or procedure sometimes • completes tasks appropriately and accurately sometimes • requires teacher guidance frequently • needs additional time, opportunities • demonstrates some Level III competencies but is inconsistent	• demonstrates some understanding of more, less, before, after, between • requires guidance in skip counting • represents numbers in limited ways • uses different counting strategies but is not consistently accurate • compares and orders sets and numerals of single-digit numbers; has difficulty with some 2-digit numbers	• uses a limited number of directional, positional, comparative words • identifies some plane and solid figures but may not recognize them in the environment • creates models of plane figures with assistance
Level I	• exhibits minimal performance • shows limited evidence of conceptual understanding and use of strategies • responds with inappropriate answer and/or procedure frequently • very often displays misunderstandings • completes tasks appropriately and accurately infrequently • needs assistance, guidance and modified instruction	• uses counting strategies • identifies, creates sets with small numbers • recognizes some numerals • identifies "one more than," "one less than" but is inconsistent	• recognizes circles • identifies likeness as by color and size • models plane figures with assistance • needs additional clues to respond to directional, positional words

Figure 1.   First Grade Observation Matrix for On-Going Assessment and End of the Year Evaluation (concluded)

Classroom assessment can improve student learning when it is purposeful and assesses not only curricular objectives but also values the learning process as well as the attitudes of the learner. The key to success, however, lies in finding the time and means to retool teachers and give them every opportunity to discover for themselves the power in this type of assessment. Teachers continually seek ways to make best use of their time and maximize the benefits of daily instruction. This comes from professional growth which needs to include reading professional materials, participating in sustained staff development, and interaction with peers. These are essential pieces to making classroom assessment the key to instructional planning. It seems so simple, yet it is often a luxury that eludes most educators. The school site may be the most practical place for teachers to find the support they need. It is a place where teachers can work together and receive immediate feed back on practices that work well, or where they can find additional help in meeting stated objectives.

Setting benchmarks and expectations for what is to be learned is crucial, but this alone will not change how students are taught or assessed. The key to systematic change is supporting teachers by having time to listen to their concerns, giving them opportunities to hear and observe new strategies, and encouraging them to try, revise, and refine both new strategies and what works best for them. If we value on-going assessment practices then we must support teachers as they seek answers to these questions: *What do I want to know? What will it look like? What do I do with it now?*

But Won't They Teach to the Test?

Ruth E. Parker
Educational Consultant

Whether on standardized tests or classroom-based assessments, essential mathematical ideas must be at the core of assessment practices. Although this seems obvious, in actual practice it proves a real challenge. Many people responsible today for designing and implementing assessment in states, districts, schools and classrooms have not had adequate opportunities to understand the heart of the discipline of mathematics nor to develop their own understandings of important mathematical ideas. Both within the educational community and in public debate there is increasing talk of the need to teach and assess clear standards, targets, or goals, yet when it comes to identifying such standards, the focus is often placed on more historical views of mathematics and on the very skills that prove more trivial in the technological world of today. If we are to talk of teaching toward clear standards or goals, great care must be taken to ensure that the goal is both mathematically important and robust. Unless we identify appropriate goals, we will find ourselves holding both students and teachers accountable to the wrong measures, thereby ensuring that they will continue to be mathematically ill-prepared for the present let alone their futures. The development of appropriate assessment tools and opportunities for educators and the public to interact around those tools can be a powerful vehicle for helping teachers, administrators, parents, and the public develop their own understandings of important mathematics

Meaningful assessments can promote informed decision making on the part of all stakeholders within an educational community -- children, teachers, parents, administrators, school board members, and the public-at-large. However, given today's political climate of increased public scrutiny of educational practices and on-going public debate over what constitutes "good mathematics," development and implementation of powerful and relevant classroom-based mathematics assessments proves to be a daunting challenge. Testing of children has become more pervasive and more "high stakes" than ever before. Most educators are aware of the severe limitations of current norm-referenced testing practices. Some are aware of the damage to children's mathematics education done by such practices. It would be a mistake, however, to assume that parents, school board members, and the public-at-large are aware of the detriments of existing tests. They need opportunities to learn, and we should view our task of providing these opportunities with utmost urgency. Unless careful attention is paid to working with and educating our public as we work to educate ourselves, any efforts to fundamentally change assessment practices are likely to fall far short of the mark or to fail outright. The mathematics education community must work to help educate the profession and the public as we build assessments based on the essential mathematics that one must know and be able to do in order to be a knowledgeable and productive citizen.

If assessment practices are to support children's development of essential mathematical understandings there must be an alignment between what is being taught and assessed in a classroom on an on-going basis and what is being measured by external assessments. There is ample evidence that if the test matters, teachers will teach to the test (Resnick & Resnick, 1991). If there are incongruities between what teachers are expected to teach and what they and children are held accountable to, teaching toward "high stakes" tests will continue to be the focus of classroom instruction in all but a few classrooms. Discussions of classroom assessment will need to take place in the context of a re-examination of current district, state, national and international assessment practices. A close alignment between classroom-based and external assessment is essential if teachers and children are to have adequate opportunities to learn and do mathematics.

Any external assessment that will meet the expectations placed on testing today is likely, in the foreseeable future, to consist of multiple types of tasks including multiple choice items of the 2 - 4 minute per item nature, open-ended items of approximately 15 to 20 minutes duration per task, investigations that might range from one to several days or even weeks, and portfolios as collections of student-selected pieces of work over time. The most useful and meaningful kinds of assessment tasks are those that ask children to engage with essential and relevant mathematics as they deepen and demonstrate their understanding of important ideas; tasks that reveal children's real understandings of mathematical ideas as well as their dispositions toward mathematics; and tasks that provide teachers, parents and other educational decision makers with information needed to make appropriate on-going instructional decisions. The mathematics community could make an important contribution to the field if they would help focus efforts on the development of a small collection of robust mathematical tasks that will promote the kind of mathematics teaching and learning that is more likely to result in informed citizens well prepared and eager to put mathematics to work in essential and relevant ways. Such performance-based assessment tasks will be appropriate for both classroom and external assessments and will provide an opportunity for full alignment of assessment and instruction around essential learning goals.

These performance-based tasks must be mathematically important and robust. With the right tasks we could give almost the very same question to 4th, 8th, and 10th graders and expect children to demonstrate more sophisticated understandings over time as they engage with the tasks. The tasks must be central to how mathematics is put to work in the world. While learning mathematics through preparing for and doing the tasks, children will also be prepared to use mathematics in their everyday lives. The mathematical ideas embedded in these tasks must be important, complex, and compelling. The tasks must be accessible to all students while not placing an upper limit on where the mathematics can be taken. They will integrate mathematical ideas from varied domains (e.g. number, data analysis, spatial relationships, measurement, algebraic thinking). We will learn what children can do by watching them as they work on these tasks over time. Tools and support materials can be developed that can help focus ourselves, our teachers, our parents and our children on the important mathematics embedded in the tasks. These assessment tasks must have the potential to be fully and authentically integrated into on-going instruction. They must be fully consistent with the idea that mathematics is a sense-making process. In doing the tasks, children will put mathematics to work in order to figure something out. Children's work on these tasks over time will provide evidence of their deepening mathematical understandings.

My thinking around this issue was provoked by the work of the New Standards Project (NSP) under the direction of Lauren Resnick at the University of Pittsburgh and Mark Tucker at the National Center for Education and the Economy. Several years ago they produced a document that identified major ways that mathematics is put to work. These include (a) empirical data based research projects around an issue of social concern, (b) empirical data based studies of a physical phenomenon, (c) design that includes two components -- both designing something within given constraints and communicating the design so that someone else could reproduce it, (d) mathematical investigations, and (e) networking or systems. The NSP document will prove useful to those of us working to identify robust mathematical investigations as assessment tasks. As we look through this list, the potential for developing the types of tasks that authentically integrate the curriculum areas of mathematics, science, social science, and language arts also is evident.

Two Sample Tasks

Task One: Data Collection and Analysis

Students at 4th, 8th, and 10th grades might be asked to generate a question that they think is important, collect data and analyze the data in order to answer their question, then present their findings to an appropriate audience. We would expect that 4th graders will be able to generate a question and collect data although they are unlikely to use very sophisticated means of data collection or analysis. They will be able to display and interpret the data through bar graphs, line plots, etc. We would expect that 8th graders would be able to deal with issues such as sensitive data, unbiased sampling, searching for trends in the data, etc. We would expect that by 8th grade they will understand that the reason for displaying the data in a variety of ways (e.g. line plots, stem and whiskers, histograms, etc.) is for the purpose of searching for trends in the data. They will be able to make appropriate decisions regarding how best to display the data in order to convey important findings. We would expect 8th graders to do a more sophisticated analysis of the data including interpretation and appropriate uses of measures of central tendency. As students progress through the grades, we will expect them to bring more sophisticated understandings of data collection and analysis to the task. There is no upper limit on where 10th grade students can take the task mathematically since statisticians and others continue to invent new methods and procedures as they search for increasingly sophisticated ways to collect, analyze, and interpret data.

We must understand a mathematical idea (and how it develops in complexity over time) in order to assess children's understanding. Currently, few teachers or administrators have had opportunities to develop their own understanding of important mathematical ideas. Thus, they are unprepared to recognize where children are in their development of these ideas. Current and yet-to-be-developed resources must help teachers and others to understand the important mathematics of data collection and analysis and to assess if progress is being made in children's development of these ideas. As we continue over time to view student work around this task, we will be better able to identify the understandings that children reveal in their work and the levels of performance and mathematical dispositions they are capable of when well prepared for the task. By observing and analyzing the work of children who have had rich opportunities to engage in data collection and analysis over time, we can build our capacity to communicate with teachers and others about the kinds of tasks and mathematics we know children are capable of. We also can provide focused support that will help teachers better understand the levels of mathematical understanding revealed in their children's work.

Task Two: Scale Modeling

Students at grades 4, 8, and 10 can build a scale model of something in their environment or perhaps something of their own design. Again, the mathematics involved in scale modeling is so robust that there would be no upper boundaries on what students can do with the task. The task will provide an opportunity to assess children's understandings of a variety of important mathematical ideas including scaling, ratio and proportional reasoning, similarity and congruence, linear, area and volume measurement, translation between two dimensional nets or blueprints, and three dimensional models.

The Role of Students

Assessment, when used most authentically, will provide students with real opportunities to engage in mathematical thinking and to examine and communicate their understandings in the process of doing mathematics. Students are central players in this type of assessment. As students learn more about what is essential in mathematics, they will develop the understandings needed to put mathematics to work in meaningful and relevant ways. Mathematics portfolios can play an important part in students" on-going reflections on their work. In preparing a portfolio, students periodically go through their work in mathematics and select a piece of work that they think is important. In selecting work for their portfolios, students have opportunities to reflect on their work while selecting a "most important" piece of work, make determinations about the quality of their work, determine how best to communicate their findings to others who will be reading and interacting with them around their portfolios, and participate in active and on-going reflective goal setting.

The use of student-selected portfolios can also provide opportunities for teachers to observe what mathematics has been important to the children as she watches them make selections. There will be opportunities to interact individually with children and collectively with the class around portfolio selections. Portfolios can provide evidence to parents of the kinds of mathematics their children are learning and of their children's performance in mathematics. They can provide useful information to teachers and administrators, as well, as they work to assess and improve the mathematics instruction offered to children. A word of caution is in order here. Use of portfolios in the primary grades should be approached with caution. We must recognize the potential of primary portfolios to push instruction toward the kinds of tasks where there is a finished product on paper. This focus on products can easily divert instruction away from the kinds of learning environments most appropriate to a young child. We must recognize that the real work of young children in mathematics takes place in the brain as they act on physical objects in their environment. A premature focus on recording can interfere with a young child's development of mathematical understandings (Richardson, 1997).

Summary

Teachers can be asked (and more importantly, will want) to use these assessments and even to teach to this test because this is a test that will better prepare students to use mathematics in their everyday lives. With the right kinds of tasks, teaching to the test is desirable because it ensures that children have on-going opportunities to interact with important mathematics. While teaching to the test, teachers and their students both deepen their understanding of important mathematical ideas and become better prepared to put mathematics to work in meaningful and relevant ways. In short, we will have developed a test that we want and expect teachers to teach to.

The development of this small collection of robust mathematical tasks has the potential to result in an assessment system that can drive instructional practices in directions appropriate to the needs of our nation as we prepare for the world of the 21st Century. If assessments provide sound and authentic encounters with meaningful mathematics, then children's work can and will inform all stakeholders within an educational community -- especially the long range planning and daily decision making of teachers and children.

References

Resnick, L. B., & Resnick, D. (1991). Assessing the thinking curriculum: New tools for educational reform. In B. R. Gifford & M. C. O'Connor (Eds.), *Changing assessments: Alternative views of aptitude achievement and instruction* (pp. 38-75). Boston, MA: Kluwer.

Richardson, K. (1997). *Math time: The learning environment.* Norman, OK: Educational Enrichment.

Man Scars -- So Seems Lost

Sid Rachlin
East Carolina University

Hidden within *classroom assessment* is the need to first understand the paradigms our students bring to the classroom. We must look at the lenses through which they see the world -- as well as at the world they see. The lens itself shades how they interpret the world. We think we see the world as it is. We don't. We project our paradigm -- frame of reference -- onto that reality. We do not see it as it is.

Our paradigms and those of our students, correct or incorrect, are the sources of our attitudes and behaviors, and ultimately our interrelationships. The more aware students are of their basic paradigms, maps or assumptions -- and the extent to which they have been influenced by their experiences -- the more they can take responsibility for these paradigms, examine them, test them against reality, listen to others, and be open to their perceptions, thereby getting a larger picture and a more objective view.

To Know What We Know

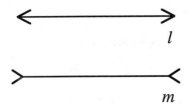

Which is longer—the length of the line segment between the arrows in line *l* or line *m*? Or are the two line segments the same length?

Did you answer that the line segments are the same length? Actually, segment shown in line *m* is shorter than one shown in line *l*. We teach illusions as if they are based on understanding. For many of our students mathematics is learned as an illusion and as with illusions understanding is based on accepting the unreasonable as reasonable.

To Understand What One Who Understands Understands

The tragedy of man is that, like the dog, his character can be molded. You cannot mold the character of a cat, an animal superior to a dog. You can give a dog a bad conscience, but you cannot give a conscience to a cat. Yet most people prefer dogs because their obedience and their flattering tail-wagging afford visible proof of the master's superiority and worth.

The nursery training is very like the kennel training; the whipped child, like the whipped puppy, grows into an obedient, inferior adult. And we train our dogs to suit our own purposes, so we train our children. In that kennel, the nursery, the human dogs must be clean; they must not bark too much; they must obey the whistle; they must feed when we think it convenient for them to feed. (Neill, 1960, p. 100)

For many, classroom assessment is asking common questions; for some, it is asking uncommon questions; and for others it is asking common questions in uncommon ways. There is a story told of an English examiner, who was invited to visit a Kindergarten classroom in which all the children had learned to conserve number as well as volume and mass. Intrigued, the examiner began interviewing each of the students in the class. One by one the children would visit with the examiner. During the interview the examiner placed two of the same style glasses before each student. Next, she poured some soda pop into one of the glasses. Then, she told the student that she wanted to pour the same amount of the soda into the empty glass. With the student's guidance, she slowly poured the soda into the glass until the student told her to stop. Next, the student confirmed that the amounts in the two glasses were the same or helped the examiner adjust the amounts until the student indicated that the amounts were the same. Now the examiner introduced a third glass that was taller and thinner than the other two. After the examiner (in full view of the student) poured the soda from the second of the two filled glasses into the taller glass, the student was asked whether there was more soda in the first glass or the tall thin glass or whether the two amounts were the same. In each case, the Kindergartners responded that the amounts in the two glasses were the same. They had *learned* to conserve. As a final step to the interview, the examiner told each student, that he or she could drink one of the glasses of soda. In every case, the student picked the soda in the taller/thinner glass. One of the challenges of classroom assessment is to separate what students know from how they have been trained to respond.

Mathematics is Learned Through Communication

"Students need opportunities not just to listen, but to speak mathematics themselves" (Silver, Kilpatrick, & Schlesinger, 1990). Although it is not uncommon for students to be asked to put homework problems on the chalkboard or the overhead projector so that other students can see their work, students are seldom asked to put into words not only what they did but also how and why they did it. This is particularly true when the tasks they are asked to explain are repetitions of trained algorithms. I am reminded of a third grade classroom that I had an opportunity to visit. The teacher was explaining how to do three-digit subtraction with regrouping.

Suppose I have 568 − 279.

$$4\,\overset{15}{\cancel{5}}\overset{1}{\cancel{6}}8$$
$$-279$$
$$\overline{289}$$

You can't subtract eight from nine, so you'll have to regroup. Eighteen subtract nine is nine. You can't subtract five from seven, so you'll have to regroup. Fifteen subtract seven is eight. And four subtract two is two. The answer is 289.

Although the teacher's arithmetic was correct, her words were not. You *can* subtract eight from nine and five from seven. After she presented another example to the class, she sent five students to the board -- each with a different subtraction problem. After all of the students completed this problem and the class had an opportunity to finish all five problems, she asked the students to explain what they had done. Each student in turn parroted the same incorrect justification. For example, if the problem was 253 − 168, the child would begin by saying "You can't subtract three from eight."

For students' communication to help them make mathematics meaningful it must go beyond recitation. Instead of a repetition of the demonstrated algorithm, students might be asked to find a three-digit number that subtracted from 253 has a difference that is greater than 60 and less than 100. By opening the problem up, you provide greater opportunities for discussion. How did you solve the problem? Why did you choose to do it that way? Did anyone solve the problem another way? Is the answer unique? Notice that each correct solution provides an opportunity to discuss subtraction with regrouping.

A Simple Example

To create a non-routine *routine* task we need only switch the given information of a standard textbook problem. For a binary operation *,
we are given that

$$O \ * \ \Delta \ R \ \square$$

where * is an operation, such as addition, subtraction, multiplication, division, raise to a power, find the root of; R is relation such as equal to, less than, or greater than; and O, Δ, and $\square$ are numbers or algebraic expressions.

Add $7 + 5 =$
becomes:

Seven added to what number equals 12	$7 + \Delta = 12$
What number added to 5 equals 12	$O + 5 = 12$
Find two numbers whose sum is 12	$O + \Delta = 12$

Find the sum,
$2x^2 - 5xy + (2xy + y^2)$
becomes:

What polynomial added to $2x^2 - 5xy$ equals $2x^2 - 3xy + y^2$?	$(2x^2 - 5xy) + \Delta = 2x^2 - 3xy + y^2$
The polynomial $2xy + y^2$ added to what polynomial equals $2x^2 - 3xy + y^2$?	$O + (2xy + y^2) = 2x^2 - 3xy + y^2$
Find two polynomials whose sum is $2x^2 - 3xy + y^2$.	$O + \Delta = 2x^2 - 3xy + y^2$

Simplify $64^{-\frac{2}{3}}$
becomes:

What number raised to the $\qquad$ $O^{-\frac{2}{3}} = \frac{1}{16}$
$-\frac{2}{3}$ power equals $\frac{1}{16}$?

64 raised to what power $\qquad$ $64^{\Delta} = \frac{1}{16}$
equals $\frac{1}{16}$?

Find an integer that raised to $\qquad$ $O^{\Delta} = \frac{1}{16}$
a power equals $\frac{1}{16}$.

The effect of the position of an unknown quantity can greatly effect a student's solution process and the level of difficulty that the problem presents. Wagner, Rachlin, and Jensen (1984) interviewed ninth grade algebra students in Athens, Georgia, and Calgary, Alberta, with a series of problems based on variations in the missing terms of the form $\square * b = c$ and $a * \square = c$ where * represents an algebraic operation and a, b, and c are whole numbers, fractions, polynomials, algebraic fractions, or radical expressions. They found wide differences in the ease with which students were able to solve these missing term tasks depending on the operation substituted for *. The difficulty of the problems varied depending on the operation and the placement of the missing term.

For example, the following missing term task was a problem for most ninth-grade algebra students.

What number multiplied by $\frac{2}{3}$ equals $\frac{3}{2}$?

Many students simply multiplied $\frac{2}{3} \times \frac{3}{2}$, while others were unsure whether to represent the problem as $\frac{3}{2} \div \frac{2}{3}$ or $\frac{2}{3} \div \frac{3}{2}$. Students tended to be rule-oriented and liked to state generalizations about the solution process before beginning the problem. In many cases students falsely generalized how to do the problem, either over-generalizing something that occurs in the problem or something they heard their teacher say (Jensen, Rachlin, & Wagner, 1982). The ninth graders who solved the problem used a variety of techniques.

Kathy solved the problem like an equation. First, she wrote $\frac{2}{3}n = \frac{3}{2}$, then she multiplied both sides of the equation by $\frac{3}{2}$.

Curtis said that first he found common denominators for the $\frac{2}{3}$ and $\frac{3}{2}$ and then rewrote the problem $\frac{4}{6} \cdot (\ ) = \frac{9}{6}$. But $6 \cdot 1 = 6$ and $4 \cdot \frac{9}{4} = 9$ so the answer was $\frac{\frac{9}{4}}{1}$.

Margaret solved the problem a third way. First she multiplied the $\frac{2}{3}$ by $\frac{3}{2}$ to get 1. Then she multiplied the 1 by $\frac{3}{2}$ to get $\frac{3}{2}$. She wrote this as $(\frac{2}{3} \cdot \frac{3}{2}) \cdot \frac{3}{2} = \frac{3}{2}$. The fraction she was looking for was $(\frac{3}{2} \cdot \frac{3}{2})$ or $\frac{9}{4}$.

Lani had still another way to solve the problem. She wrote the $\frac{3}{2}$ as an equivalent fraction such that 2 would divide into its numerator evenly and 3 would divide into its denominator evenly. She chose $\frac{18}{12}$ as a fraction equivalent to $\frac{3}{2}$ that met the conditions. Then $\frac{2}{3} \cdot ? = \frac{18}{12}$. Since $18 \div 2 = 9$ and $12 \div 3 = 4$, she decided her solution was $\frac{9}{4}$.

Joe solved it a fifth way. First he wrote $\frac{2}{3} \cdot (\ ) = \frac{3}{2}$ and then he filled in the parentheses. Since he wanted to get a 2 in the denominator, he needed a 2 in the denominator of the fraction in the parentheses. But that 2 would "cancel" with the 2 in the numerator of the $\frac{2}{3}$ so he needed a 4 in the denominator of the fraction in the parentheses. Similarly, he said that he needed a 9 in the numerator.

By changing the problem from "Multiply $\frac{2}{3} \times \frac{9}{4}$." to "What number multiplied by $\frac{2}{3}$ equals $\frac{3}{2}$?" the nature of the variety of ways that students solved the problem changed. Each student's method is rich with opportunities for discussion. Students' errors provide opportunities for new directions or to clarify misconceptions. For example, Curtis' complex fraction solution opened up a new topic for discussion. Although the problem is not much different than the original task, the richness of the discussions is quite different.

How students perceive a problem shapes the other processes which they may bring to bear on the solution of the problem. The various solution paths which students select establish their structure for the problem. For example, the problem "What number divided by 24 equals $\frac{3}{4}$?" has a wide variety of appropriate solution paths depending on the way in which the task is perceived, for example, as equivalent fractions, a proportion, a division problem, or an equation. Students who have only one way to see a problem limit their ability to solve it and to link it to other mathematics. If a student tells you that they solved the problem like equivalent equations, what equation do you think they wrote to represent the problem? What if they said that they solved it like a proportion? Suppose instead of using words, I had given you the problem with a blank to be filled in, $\square \div 24 = \frac{3}{4}$, would you think of solving the task like equivalent fractions? By providing the problems in words, the students are given an opportunity to translate the problem into their own symbolic or pictorial representation. Each representation is rich with its own suggestive solution paths based on a student's past experience.

$$\frac{\square}{24} = \frac{3}{4} \qquad\qquad 24\overline{)?}^{\frac{3}{4}}$$

We See the World Not as It Is, but as We Are.

What is the relationship between a student's structure for problems and his or her teacher's anticipation of the student's structure? Wagner, Rachlin, and Jensen supplemented their study of students' learning difficulties in elementary algebra with the classroom teacher's analysis of the students' problem-solving processes in algebra. After eight interviews were conducted with each of ten students in Athens, Georgia, and four students in Calgary, Alberta, the classroom teachers were asked to complete all of the interview tasks. Then they were requested to guess how each of their students would solve the problems. Finally, the teachers were able to listen to (in Georgia) or watch (in Alberta) the interviews to test the accuracy of their predictions. Rachlin (1982) reported on results of the interviews with the Calgary teacher. The teacher was very flexible with tasks such as "What number added to the sum of 17 and 6 equals 6?" and could solve the tasks in several ways. But, he was surprised to find that only one student in his advanced section solved this problem by noting the sixes in both members. Instead the others first added 17 and 6 and then subtracted 23 from 6.

The structure of the students' solutions, at times, varied from classroom practice. For example, the teacher was surprised that three of the four students solved tasks such as "What trinomial subtracted from $5x^2y - xy^2 + 7$ equals $-x^2y + 8$?" by writing the parts vertically. With regards to the vertical form, the teacher commented, "They've seen it occasionally in the textbooks, but I've never assigned the problems that have vertical form." At the same time that students create their own approaches, they may not realize the relationships between the approaches they do not own. For the child, there may be no reason to assume that two different ways to solve a given problem should be connected. In the following account, my son Jeff was just beginning second grade.

Beside Jeff's bed hung a chalkboard. If I had been a phys. ed. teacher, I probably would have placed a trampoline there, but as a math teacher a chalkboard won out. Every so often I would come into his room and write a math problem on his board. At another time, he'd come in and solve the problem. Still later, I'd stop back and check his solution. If it was correct I'd erase it and place another problem on the board. If it was incorrect I'd call him in and we'd discuss it.

On one occasion, when Jeff had just been introduced to addition with regrouping (carrying), I wrote the problem:

$$\begin{array}{r} 24 \\ +16 \\ \hline \end{array}$$

When I later returned to the room, I found Jeff's solution:

$$\begin{array}{r} 1 \\ 24 \\ +16 \\ \hline 41 \end{array}$$

After I called Jeff into the room, the following dialogue ensued.

"Jeff, I think there's an error here."

"No, there isn't Dad. Look! Four and six are ten. You put down the one and carry the one. One and two are three and one is four. The answer is 41."

"No, Jeff, I think something is wrong here."

"Look! [Jeff spoke a little louder to make his explanation clearly more acceptable.] Four and six are ten. You put down the one and carry the one. One and two are three and one is four. The answer is 41. You ask Miss Frame, she'll tell you how to do these."

As a math teacher I took this as a sign that a concrete embodiment was needed. After all, using concrete objects makes math make sense. I left to get a pack of toothpicks and a box of rubber bands and returned to sit on the floor beneath the chalkboard. Jeff had grouped by tens before and had no difficulty representing 24 as two tens and four ones and 16 as one ten and six ones. He added (combined the two piles) and got three tens and ten ones or after trading in the ten ones for one ten he had an answer of 40 with the toothpicks. At this point he looked back and forth at his pile of toothpicks and the chalkboard. Finally, he said very seriously, "That's what you get when you add toothpicks, but when you work on the board you get this answer."

To my surprise, I later learned that Jeff's response is not that unusual. It has been reported by other parents and teachers (Wirtz & Kahn, 1982). But nonetheless, it serves as an important reminder. Why should children expect that what they get on the board should match what they get with objects? If we are using concrete objects to serve as a foundation for arithmetic operations, we must make sure that the procedures used with the concrete materials parallels the procedures used in the rote algorithms being taught. The multiple representations for problems must be linked within the students' minds, as well as within the teachers'.

Seek First to Understand -- Then to be Understood

In coming to understand, we are first struck by the errors our students make -- the scars of their past experiences. Our first instinct is try to treat the students as tabula rosa, hoping to fill their clear surface with meaningful understanding. Like writing on a well-worn chalkboard, these efforts are often exacerbated by the prior etchings in the slate. Our next instinct may be to view the scars as impediments towards learning -- that a child, once scarred cannot be healed. The challenge of classroom assessment is to help us shift our paradigms. Rather than ignore the scars or give up because of their existence; we must learn to view the he scars as steps towards a solution.

The goal of classroom assessment is to inform. When successful, it guides instruction and facilitates the development of understanding. The title of this paper is more than an anagram for *classroom assessment*. It provides a cryptic reminder that before we try to have students understand, we must try to understand the paradigms our students bring to the classroom. If we rearrange the letters in *classroom assessment* we find the message *Man Scars -- So Seems Lost.*

> *Before our students will understand, we must first understand our students:*
> - *We must understand what one who understands understands.*
> - *We must understand what one who does not understand understands.*
> - *We must understand what one who teaches one who does not understand to understand understands.*
> - *We must understand what curriculum will lead one who does not understand to understand.*
> - *We must understand what curriculum will lead one who does not teach for understanding to teach for understanding.*

References

Jensen, R., Rachlin, S. L., & Wagner S. (1982). *A clinical investigation of learning difficulties in elementary algebra: An interim report.* Paper presented at the Special Interest Group/ Research in Mathematics Education and the Research Advisory Council of the

National Council of Teachers of Mathematics joint meeting, Toronto, Ontario, Canada.

Neill, A. S. (1960). *Summerhill: A radical approach to child rearing.* New York, NY: Hart Publishing.

Rachlin, S. L. (1982). A teacher's analysis of students' problem-solving processes in algebra. In S. Wagner (Ed.), *Proceedings of the fourth annual meeting of the North American Chapter of the International Group for the Psychology of Mathematics Education* (pp. 140-147). Athens, GA: University of Georgia, Department of Mathematics Education.

Silver, E. A., Kilpatrick, J., & Schlesinger, B. (1990). *Thinking through mathematics: Fostering inquiry and communication in mathematics classrooms.* New York, NY: College Entrance Examination Board.

Wagner, S., Rachlin, S. L., & Jensen, R. J. (1984). *Algebra Learning Project final report.* Athens, GA: University of Georgia Department of Mathematics Education.

Wirtz, R. W., & Kahn, E. (1982). Another look at applications in elementary school mathematics. *Arithmetic Teacher, 30*(1), 21-25.

Assessing Core Concepts in K-2 Mathematics

Kathy Richardson
Mathematical Perspectives

Across the nation, teachers are working hard to improve the teaching and learning of mathematics. Teachers want the mathematics they present to their students to be meaningful and to be aligned with the Standards (National Council of Teachers of Mathematics, 1995) which challenge teachers to broaden their student's mathematical experiences. Growing numbers of teachers realize that the learning of mathematics requires more than memorizing rote procedures. They know that children must make sense of the mathematics they are learning if they are to use mathematics to analyze situations and to solve problems. Teachers of children in Kindergarten through Grade Two have responded to the challenge presented by the *Standards* by involving children in activities designed to bring meaning to mathematical concepts.

Many teachers, however, who are working hard to bring meaningful mathematics to their students find themselves uncertain and even anxious about assessment. In the past when teachers followed textbooks page by page, they thought their charge was to see that the children listened, followed directions, worked hard, and could do the problems on the page. Now, many of these teachers see teaching mathematics as a process of providing interesting activities rather than completing workbook pages. Yet, it is not always clear to those teachers using new approaches and new materials what the children should be learning from the activities or what to look for to see if learning is occurring. They focus on making sure the children do the activities, trusting that if they present what they believe are meaningful activities, the children will learn.

Using activities to teach for understanding is a more complex process than teaching procedures and poses different problems for teachers. When the goal is learning procedures, the teacher can easily tell whether or not the child has learned to do the particular procedure. However, when the goal is understanding, teachers often don't know when the activities have been effective and when they have not. Knowing that children must construct understanding of mathematical ideas for themselves, teachers aren't always sure when to intervene and when to just let children "do the best they can." Sometimes, teachers' uncertainty about whether or not children are learning what they are supposed to be learning shows up in questions and comments such as the following:

- We've done all these activities, but my children are still counting by ones instead of by tens. How long do I wait before I show them how?
- We did a whole unit on multiplication, but the kids still don't know their multiplication facts.
- My students really love all these great math activities, but we just can't spend all our time on them. I have to get the kids ready for next year, too.

Teachers need to be able to determine what the children are learning or not learning from the activities they give them to do. They need ways to find out if children are growing in their understanding of concepts. To this end, some primary teachers are being asked to use alternative assessments in the form of portfolios and performance tasks with accompanying rubrics. These alternative forms of assessment, which can be valuable in the intermediate classroom, do not always translate directly into primary classrooms. When portfolios are required to include mainly children's written work, primary teachers often find it necessary to begin a search to find those activities that will result in products suitable for inclusion in the portfolios. The mathematics program then becomes product oriented rather than focused on the essential mathematical concepts and understandings that children must acquire. Many of the rubrics developed to go with performance tasks deal, not with whether children are developing concepts, but rather, with how well a child completed that particular task. Oftentimes the rubrics are not specific enough about what mathematics is being dealt with in the task. Thus, they are not helpful to teachers wanting to interpret the information in order to make appropriate instructional decisions.

Those of us who teach primary-aged children, who are just learning to read and write, must consider carefully what these kinds of assessments can and cannot tell us about our young children's development of mathematics concepts. What primary children are able to produce on paper often gives more information about their ability to write than the level of their mathematical thinking or growth in understanding of mathematical concepts. It is true that children need experiences which require them to write about the mathematics they are doing. We can even get some hints about their mathematical thinking through this work. However, in the primary grades, children's written work cannot provide definitive information about their current levels of mathematical understanding.

Teachers are feeling the pressure to use portfolios and performance tasks. At the same time they are also struggling with conflicting pressures and admonishments not to "throw the baby out with the bath water." They are reminded that they must also be accountable for making sure children know the basic skills. Some teachers handle this by separating skill development from mathematical activities and they end up with two kinds of assessments: 1) a portfolio showing children's work with activities and 2) traditional paper and pencil tests used to determine if children are learning the basics. Too often neither of these kinds of assessments provides the teacher with the critical information they need to plan a mathematics program that supports their student's developing understanding of mathematics concepts.

We see, then, many teachers who are eager and willing to do the right thing. But they are left with conflicting messages which makes it very difficult for them to figure out what is most important for their students.

What Teachers Need to Know

Teachers need to know how to use assessments that will help them determine not how well children follow procedures or complete tasks, but rather how well they *understand* mathematical ideas and concepts. They need to be able to recognize the stages that children go through when developing understanding and competence and take this into consideration when deciding what kinds of experiences the children need.

If teachers are going to become better able to assess the development of mathematical understanding and to make sound instructional decisions based on these assessments, they must have opportunities to grow in their knowledge of certain principles that impact the way teaching and learning mathematics is experienced in the classroom. The following are some of the suppositions related to assessing for understanding that teachers could consider.

Making Sense of Ideas and Situations

All the current efforts to improve the teaching of mathematics emphasize that children must understand and make sense of mathematics if they are to become truly competent mathematically. In order to make sense of the mathematics they are learning, children need to stop reading the teacher's face for clues to right answers. Rather, they need experiences that help them learn to trust their own thinking and ability to find out. Children are not making sense of mathematics when they learn basic concepts by rote or learn to follow procedures that have little or no meaning for them. Making sense cannot be relegated only to problem solving experiences but must be an essential part of the learning of basic skills as well.

No matter what mathematics they are learning, children need to confront the following types of questions:

- Does it make sense?
- Will it work every time?
- How do you know?
- Are you sure?
- How can you find out?

The challenge for teachers is to find ways to maximize children's learning without interfering with their sense making process.

Understanding and Getting Right Answers

We often make assumptions that children understand what they are doing if they are able to get right answers. However, we need to assess more than the children's answers and uncover the thinking behind the answers. If we are to consider an assessment to be authentic, we must make sure it elicits authentic responses and helps the teacher determine what children really know and understand, not just what they can do or say. If children can be successful with a task without understanding, it is not a good enough assessment.

Assessment as a Guide to Instruction

Classroom assessments should be much more than a way for teachers to document what a child is doing in class. Ongoing assessments of children's developing mathematics concepts should directly influence the instructional decisions the teacher makes. The assessments have value if and when the resulting information helps the teachers better meet the varying needs of their students.

Core Concepts

Problem solving and mathematical thinking are important aspects of the mathematics program, but there are also important core concepts that children must learn if they are to be powerful mathematically. Ideas that may seem trivial to the advanced mathematician are actually the important intellectual work of the young child. Children need to experience these concepts in a way that honors the complexity of these ideas and encourages thinking and making sense. These important core concepts must be identified and made clear to the teacher.

Stages of Thinking and Understanding

Since children move through many different stages as they develop competence mathematically, teachers need assessments that help them look at the various levels of children's developing understanding. The assessments that help identify various stages are important because understanding and competence do not happen on the same time line for all children. When teachers have the information they need, the instructional activities will not be ends in themselves, but rather tools to develop mathematical competence.

Assessing Children's Understanding in the Classroom

Teachers need practical and simple assessment tasks that help them identify what children already know as well as what they need to learn. In order to assess for understanding, teachers need to know how to create a learning environment that allows for authentic individual responses and opportunities for observations by the teacher.

Choosing Concepts to Assess

The developing mind of a child is far too complex and ever-changing to be easily analyzed in terms of a check list. Teachers could never keep track of all that their children are learning even in one subject area such as mathematics. Rather than trying to keep track of too much, teachers and children will be better served if teachers are provided with assessments that help them focus on what is most important for discerning a child's developing understanding of mathematical concepts. The challenge is to find reasonable and manageable ways to obtain and record critical information about children's mathematical learning. This can be accomplished if certain core mathematical ideas, vital to the ongoing development of mathematical thinking and understanding, are identified.

There are many core concepts that span the range of mathematical topics that children should experience in elementary school. However, we should not underestimate how important it is for children to build a strong foundation in number. What children know and understand about number impacts their work with all other mathematical topics. Therefore, we must keep the development of number concepts at the heart of the mathematics program for young children. Children need ongoing and multiple opportunities to develop number sense: to count and compare quantities, to add and subtract, and to work with place value in ways that ask them to think and reason, to see relationships and to make connections. So, for this publication, I have chosen what I believe to be a core concept in number for each grade level: kindergarten through second grade. These core concepts develop over the course of many months and will, certainly, develop at different times for different children. However, it is also true that most children will benefit from working with these ideas within a predictable time span. I have indicated the grade level that would typically be the time that most children will benefit from work with the particular idea described.

Initial Number Concepts (Counting to 10 with Extensions to 20 or 30)

The most important foundational work of the young child is learning to count while simultaneously developing a sense of number and number relationships. Children learn the language of counting long before they learn what the numbers represent. It takes a long time to develop a sense of quantities and relationships. When we carefully observe young children counting objects, we see how challenging the task can be for them. When children first learn to count, they are focused on saying the right words and landing at the right place. At first they don't even realize they should end up with the same number every time they count the same pile of objects. It is a sign of progress when we see them count and recount in order to make sure they end up with the correct number. We can also see the young child's sense of quantity develop when we ask them to estimate. In the beginning we will hear them say what seems to be any number that comes to mind. The ability to make reasonable estimates improves when they are given experiences that ask them to think about the quantities with which they are working.

Basic Addition and Subtraction (Facts to 10 with Extensions to 20)

It has long been recognized that knowing the basic facts is essential to later work with mathematics. However what it means to know the basic facts is a subject worth examining. Children who can use basic facts to work with more complex problems understand that they are describing number relationships. They can take numbers apart and put them back together again. They are able to think with these relationships rather than simply parroting answers back as though they were memorized rhymes. They know these basic relationships so well they do not have to figure them out or recreate them in their minds. The process of learning the combinations for numbers to ten is a much slower process than many of us realized in the past. Both kindergarten and first grade children are capable of memorizing "facts" but these facts are often not useful tools for them. This is another example of the fact that children can learn the language of numbers before they fully understand what this language is describing.

Seeing the smaller numbers that make up larger numbers and describing those number parts is the first step in learning basic facts. Many children in kindergarten are at a stage of thinking where they don't yet see the smaller numbers in the larger numbers. If you were to ask them to show you the three and the two in five, many would say, "That's not three and two. That's five. See. One, two, three, four, five." Many first grade children are just becoming aware of the little numbers that go together to make bigger numbers and still need to count to find out how many altogether when combining the parts. Others are beginning to know combinations such as 3 and 3 make six and 5 and 2 make seven. Of course, it is true that some kindergarten children do know these basic relationships and some second graders do not. We can't assume children know or do not know these relationships without assessments to see what they are able to do in a variety of situations.

Place Value (Tens & Ones with Extensions to Hundreds, Tens, Ones)

Teachers of older children know how often children struggle with place value concepts. The lack of understanding shown by 4th and 5th graders does not occur because they haven't worked with place value ideas in the past. Rather, it often happens because children have been asked to work with place value concepts before they are ready to understand. Too often we haven't recognized the complexity of these ideas and have assumed children are learning place value concepts simply because they can use the language of tens and ones.

Place value concepts are often introduced as young as kindergarten. However, there is evidence that most of these young children do not understand what they are working with. For example, kindergarten classes commonly mark each day of school with a straw or a popsicle stick and make bundles of ten as the straws or sticks are accumulated. Kindergarten children can learn to count along with the teacher and the rest of the class and can repeat the right number of tens and ones when the teacher asks them to. However, if you ask these same children at the end of the year how many they think are in a bundle of straws, you will see that most of the children have no idea. In first grade it is common for children to organize groups into tens and ones if instructed to do so by the teachers. But when the children really want to be sure of the number of objects, they push them back together and count them all by ones. It is in second grade that most children can begin to count groups and to begin to think of numbers as composed of groups of tens and leftovers.

Ways Teachers Can Assess

While learning how to write and record experiences is important work for the young child, we know that what they understand about mathematics is not generally going to appear on paper. Therefore paper and pencil tests will naturally be inadequate if we are to find out what young children know and understand. We will get much more complete and useful information if we watch and interact with the children while they are doing mathematical tasks. There are two ways that we can get information about the child's level of understanding and competence with concepts: through observing children while they are doing their work and through individual interviews.

Observing Children at Work

Much of the most useful information teachers get about children's thinking is inferred from the actions and comments they make while they are engaged in mathematical tasks. The nature of the tasks the children are engaged in will greatly influence the kind of information that can be obtained. Tasks that can be approached in a variety of ways are much more revealing of a child's thinking than tasks that must be done in particular ways. Teachers will get more authentic information about how children think if the children are encouraged to make sense of their work for themselves instead of simply following the teacher's directions or reading the teacher's face for a clue to the answer.

It is helpful for teachers to recognize that they cannot observe children well if they try to see too many children at one time. Observations are much more effective if the teacher takes a few minutes to watch one or two children at a time rather than assuming they should be able to look at a whole group at once.

Individual Interviews
The thought of doing individual assessments is overwhelming to many teachers, but those who have taken the time to present individual assessments to children quickly realize the value of the time it takes. When a teacher observes a child after an individual assessment has been given, the teacher is able to focus in on the child and see more than he or she otherwise could. It becomes relatively easy to add to the information obtained during the individual assessment. It is extremely beneficial for teachers to take the time at the beginning of the school year to do individual assessments as they are then better able to provide appropriate experiences that will aid children in developing mathematical understanding and competence all year long.

At first, teachers sometimes find that giving individual assessments is somewhat cumbersome and overly time-consuming. However, teachers also find that with practice, the assessments soon become easy to give and end up taking much less time than in the beginning when teachers were first learning to give them.

Teachers have found many different ways to accommodate their need to work with individual students. Sometimes they assess while the other children explore mathematics materials, do quiet work such as reading or drawing, or have an assistant or parent helper read to them. Once the teacher has found a way to get this valuable information about each children, he or she sees that nothing else that they might provide for the class is more important to the instructional program as a whole during year than these assessments.

Conclusions

When teaching children particular mathematical ideas and concepts, it is important for teachers to know what they want the children to know, understand, and do. They need to know what they are looking for as indicators of developing understanding or continuing needs. With this information, they will be able to provide appropriate mathematical experiences for their students. The more teachers know the natural stages of development of mathematics concepts and the more they know what to look for while the children are working, the more they will be able to respond right on the spot with the next question and with appropriate support and challenges. Then, truly, classroom assessment and instruction will be indistinguishable.

What do Teachers and Students Obtain from Assessment of Student Work?

Judith T. Sowder
San Diego State University

Clarke (1996) has noted that assessment has three distinct and fundamental purposes: to model, to monitor, and to inform. I want here to focus on the third purpose and to discuss this purpose in terms of feedback. In order to contextualize this discussion in terms of the conference goals, I should first say that of the possible questions given to participants to discuss, two are relevant here. They are (with some changes in wording):

- How can teachers learn to collect, organize, and interpret the kinds of information they need to assess student learning, provided that the assessment tasks have already been decided upon?
- What constitutes quality feedback and in what ways can teachers provide useful feedback to students and their families?

If assessment practices at the classroom level are successful, they can inform the teacher, the parents, and the students themselves about what mathematics has been learned, and about what the next steps should be.

In writing about assessment, Clarke (1996) also distinguished between the activity of coding information and the actual process of assessment. Assigning a grade is one way of coding, but little information is provided by a single letter grade because so much complex information has been condensed or discarded in the process of assigning the grade. "Such a grading process can sacrifice precisely that detail which might contribute most constructively to the subsequent actions of teacher, student, and parent" (p. 354). There seems to be general agreement that when tasks call for complex responses from students, a scoring rubric must be devised and applied (Mathematical Sciences Education Board, 1993) that is appropriate for the task at hand.

Procedural Knowledge

1 point if some work is done but with incorrect use of procedures and operations or with major errors. No indication appears of the student's knowing reasons for procedures.

2 points for appropriate use of procedures with minor errors, and some ability to represent or explain.

3 points for correct, error-free use of procedures with correct explanations

4 points for extended use of procedures, for different explanations or reasons, for extending, adapting, or inventing new procedures.

Conceptual Understanding

1 point if there are wide gaps in conceptual understanding, major misconceptions, little or no use of terminology, diagrams, or symbols.

2 points if there are gaps in conceptual understanding and some evidence of misconceptions, but attempts are made with models, diagrams, or symbols.

3 points for good evidence of conceptual understanding.

4 points if there is, in addition to evidence of understanding, accurate use of models, diagrams, and symbols with good transitions from one mode to another.

Problem solving

1 point for an unworkable approach, incorrect or no use of mathematical representation, lack of understanding.

2 points for appropriate approach, estimation, and implementation of a strategy.

3 points for workable approach, estimation, and implementation of a strategy.

4 points for an efficient/sophisticated approach; extensive use of mathematical representations, solutions with connections, synthesis, or abstraction.

Figure 1. Anaholistic Scoring Rubric

Clarke suggests two alternatives for evaluating complex performances, both of which provide a numerical score when such a score is desirable: detailed scoring schemes, and holistic grading. In both cases, the value of the scoring schemes is that they can be shared, and can provide information on expectations. In a detailed scoring scheme, weighted numerical scores are assigned to particular aspects of a student's performance, depending on the assessor's estimate of the demands and significance of the components. A similar scheme is described by Kulm (1994) and is called "anaholistic." The distinction is that while one generalized rubric is used for holistic scoring, more than one criterion is used in anaholistic scoring, but with the possibility of summing up the scores to obtain one score. Kulm provides an example of anaholistic scoring that is based on a scoring scheme used in Oregon. Notice that there are three criteria used (Figure 1). (I have modified this example a bit.)

From such a scheme, some sort of synthesis is next needed if the information is to be useful to the teacher, to the student, or to the parent. That is, once the coding is complete, what can be said about the progress of this individual student? Of course, part of the synthesis will be individualized, based on past performance. Thus, two students receiving the same coding might have different summaries written about their work.

6 points: Exemplary Response In addition to points from competent response, gives a complete response with a clear, coherent, unambiguous, and elegant explanation; includes a clear and simplified diagram; identifies *all* the important elements of the problem; may include examples and counterexamples.
5 points: Competent Response Gives a fairly complete response with reasonably clear explanations; may include an appropriate diagram; communicates effectively to the identified audience; shows understanding of the problem's mathematical ideas and processes; identifies the most important elements of the problems; presents solid supporting arguments.
4 points: Minor Flaws but Satisfactory Completes the problem satisfactorily, the explanation may be muddled in places; argumentation may be incomplete; diagram may be not quite appropriate; understands the underlying mathematical ideas and uses mathematical ideas effectively.
3 points: Serious Flaws but Nearly Satisfactory Begins the problem appropriately but may fail to complete or may omit significant parts of the problem; may fail to show full understanding of mathematical ideas or processes; may make major computational errors; may misuse or fail to use mathematical terms; response may reflect an inappropriate strategy for solving the problem.
2 points: Begins, but Fails to Complete Problem Explanation is not understandable; diagram may be unclear; shows no understanding of the problem situation; may make major computational errors.
1 point: Unable to Begin Effectively Words do not reflect the problem; drawings misrepresent the problem situation; fails to indicate which information is appropriate to the problem.
0 points: No Attempt

Figure 2. Rubric for Open-ended Responses

There are several holistic scoring schemes that have been circulated in recent years. The example in Figure 2 is from a rubric for evaluating responses to open-ended questions (EQUALS and California Mathematics Council, 1989).

Scoring rubrics provide information to the instructor that can be used to assess the understanding and performance of students. A series of 3's and 4's using the above rubric would tell the teacher something about the level of performance of a student over several tasks, or of a class on one task. A scoring rubric can also be used to provide feedback to students and to parents if they are given a copy of the rubric. It is important that students have access to the standards by which their work is being measured if assessment is going to lead to any improvements in their work. Another way of helping students reflect on and improve the quality of their work is to provide them with their own version of the rubric being used in scoring. For example, Mathematical Sciences Education Board (1993, p. 75) provides a rubric in which the directions for students tell them exactly what four areas the reader will look for:

- How well you understand the problem and the kind of math you use
- How well you can correctly use mathematics
- How well you can use problem-solving strategies and good reasoning
- How well you can communicate your mathematical ideas and your solution

Students are then given a check-off guide for judging the quality of their responses. For example, the first area is conceptual understanding of the problem, and the check list under this point includes "I used diagrams, pictures, and symbols to explain my work, I used all the important information to correctly solve the problem," and "I have thought about the problem carefully and feel as if I know what I'm talking about." This type of rubric can be discussed by the class before being used by the students as they work on a task.

Providing students and parents with scores and/or summary accounts is but one form of feedback. Ideally, the feedback provided to students will help them to "self-assess and self-correct" (Wiggins, 1993, p. 183). Wiggins distinguishes between feedback and guidance in the following way: "Guidance gives direction; feedback tells me whether I am on course" (p. 184). His analogy was that a map or triptych might provide guidance on a trip, but landmarks and road signs provide feedback. He claims that there is a complete absence of adequate performance "road signs" in most classes, and that this constant failure to receive good feedback is a legacy of "defining education as 'teaching' and assessment as 'testing' after teaching" (p. 187). So what *is* good feedback? Wiggins (pp. 198-199) distinguishes between good feedback and poor feedback (Figure 3).

Effective Feedback	Ineffective Feedback
Provides guidance and confirming (or disconfirming) evidence.	Provides praise or blame, and non-specific advice or exhortations.
Compares current performance and trends against successful result.	Naively assumes that instruction and hard work will bring people to their goal.
Is timely and immediately useful.	Is not timely; suffers from excessive delay in usability or arrives too late to use at all.
Measures in terms of absolute progress; assesses the accomplishment; specifies degree of conformance with the exemplar, goal, or standard.	Measures in terms of relative change or growth; assesses student behaviors or attitudes; relative to the self or norms, tell students how far they have come (not how far they have left)
Is characterized by descriptive language.	Is characterized by evaluative or comparative language
Is useful to a novice; offers specific, salient diagnoses and prescriptions for each mistake.	Is not useful to a novice; offers only a general summary of strengths and weaknesses; criticism uses language only experts can "decode"
Allows the performer to perceive a specific, tangible effect of his or her efforts, symbolized by an apt score.	Obscures any tangible effect so that none (beyond a score) is visible to the performer
Measures essential traits	Measures only easy-to-score variables.
Derives the result sought from analysis of exit-level or adult accomplishment.	Derives the result sought from an arbitrarily mandated or simplistic goal statement.

Figure 3. Effective and Ineffective Feedback

The *Assessment Standards* (NCTM, 1995) provides a short paragraph on feedback which is basically in agreement with Wiggins' chart: "The best feedback is descriptive, specific, relevant, timely, and encouraging. It is immediately usable. The feedback may be oral or written, formal or

informal, private or public, geared toward and individual or a group....
Providing effective feedback in a continual and recursive manner will help
each student become an independent learner" (p. 34). An example is given
of a student's work, with the teacher's feedback comment: "Stewart, your
work indicates that you know *special* odd numbers that sum to 16, 1 + 3
+ 5 + 7, not just any odd numbers (e.g., 11 + 5). You need to be more
convincing that your pattern will *always* work" (p. 35).

There is some controversy about whether students should see evidence
of exemplary work, that is, the performance of someone achieving the
standard set by the scoring rubric. Some teachers believe that seeing
examples of what is considered exemplary work will stifle the creativity
of the students, and that they will simply try to imitate that work. If to
imitate means to provide evidence of understanding through clear,
coherent, unambiguous explanations with accompanying diagrams, then I
think we all like the idea of imitation. The fear, however, is that students
will want their own solution to look like the desired solution of the same
problem. It might, but if the next task is quite different, imitation of
format and other irrelevant details will not lead to exemplary work.
Students can learn a great deal from studying the success of others.

Appropriate feedback is the means by which students benefit from
assessment. Clarke asks the question, "What purpose does assessment
serve for our students?" and then provides the answers a student might
give:

-- 'Assessment lets me show what I know',
-- 'Assessment validates my performance and my learning',
-- 'Assessment tells me what I need to work on',
-- 'Assessment makes me accountable',
-- 'Assessment tells me what performances are valued',
-- 'Assessment tells me what is quality work in this subject', and
     even,
-- 'Assessment gives me a say in my education' (p. 346)

One major objective of the reform movement in mathematics
education is to help students take responsibility for their own learning.
An absolute prerequisite for attaining this goal is that students be provided
feedback by which they can assess their own progress.

References

Clarke, D. (1996). Assessment. In A. Bishop, K. Clements, C. Keitel, J. Kilpatrick, & C. Laborde (Eds.), *International handbook of mathematics education, Part 1* (pp. 327-370). Dordrecht, The Netherlands: Kluwer.

EQUALS and California Mathematics Council. (1989). *Assessment alternatives in mathematics.* Berkeley, CA: Lawrence Hall of Science.

Kulm, G. (1994). *Mathematics assessment: What works in the classroom.* San Francisco, CA: Jossey-Bass.

Mathematical Sciences Education Board. (1993). *Measuring what counts: A conceptual guide for mathematics assessment.* Washington, DC: Author.

National Council of Teachers of Mathematics. (1995). *Assessment standards for school mathematics.* Reston, VA: Author.

Wiggins, G. P. (1993). *Assessing student performance: Exploring the purpose and limits of testing.* San Francisco, CA: Jossey-Bass.

The Role of Mathematical Proof in Classroom Assessment

Diane M. Spresser
National Science Foundation

Over the past year, there has been a marked increase in public attention to student assessment in K-12 mathematics. The recent release of results from the Third International Mathematics and Science Study (TIMSS) has sparked great interest, both from the general public and from the university and K-12 mathematical communities. TIMSS is a large-scale, cross-national comparative study of the educational systems of some fifty nations and "their outputs" (Schmidt, McKnight, & Raizen, 1996, p. 4). The TIMSS study includes analyses of curricula, instructional practices, school and social factors, and textbooks in mathematics and the sciences, as well as the results of achievement testing of students and, from a few countries, videotapes of actual classes (Schmidt, McKnight, & Raizen, 1996, p. 4).

In the sections that follow, we consider selected results from TIMSS as a context for discussion of the importance of deductive reasoning and mathematical proof, classroom questioning strategies, assessment of student understanding, a classroom example, and some important caveats.

TIMSS Results

Among the forty-one countries on which data are reported in TIMSS, U.S. eighth-graders were above average in science and below average in mathematics. TIMSS reported achievement among seventh- and eighth-graders in six content areas of mathematics (Beaton, Mullis, Martin, Gonzalez, Kelly, & Smith, 1996, p. 12): fractions and number sense; measurement; proportionality; data representation, analysis, and probability; geometry; and algebra. U.S. students scored at about the international average in three areas of mathematics (fractions and number sense; algebra; and data representation, analysis, and probability), but did less well in the remaining three (U.S. Department of Education, 1996, pp. 27-29). In geometry and measurement especially, the performance of U.S. eighth-graders was below that of most of their international peers (pp. 28-29).

The eighth grade results of TIMSS lend support to some popularly held opinions but contradict others. U.S. eighth-graders, for example, have as many hours of instruction in mathematics per week as their peers in other countries (Beaton et al., 1996, pp. 144-147) and spend about as much time per day out of school studying mathematics (pp. 111-112). U.S. under-performance in mathematics therefore appears to be related to *how* that time is spent, rather than to the *amount* of time invested. The TIMSS videotape component, which includes tapes of U. S., German, and Japanese middle grades mathematics classrooms, offers some interesting insights, especially since Japanese students significantly outperformed both U.S. and German students on the TIMSS mathematics achievement test. The national overall mean achievement score in mathematics for German students is also higher than for U.S. students, although not significantly so (U.S. Department of Education, 1996, pp. 19 - 20). The video segments suggest, for example, that more time is spent in U.S. classrooms going over homework assignments from the previous day and getting started on newly assigned homework. Mathematics lessons in U.S. classrooms are also subjected more frequently to unrelated interruptions, such as a loudspeaker announcement or a class discussion of a recent sports event (U.S. Department of Education, 1996, pp. 43 - 44). In both Japan and Germany, more than in the U. S., mathematics lessons tend to be comprised of a single component and emphasize multi-step tasks. In other words, lesson "segments" in Japan and Germany are more often connected or explicitly linked, resulting in greater coherence. Mathematics lessons in the U.S. have less emphasis on reasoning than in Germany and especially Japan (Stigler & Manaster, 1997). In the TIMSS videotape research, "the lowest rating was assigned to the mathematical reasoning used in 87 percent of the U.S. lessons, in comparison to 40 percent of the German and 13 percent of Japanese lessons" (U.S. Department of Education, 1996, p. 45).

While the TIMSS results in mathematics for U.S. eighth-graders have been disappointing, the TIMSS fourth-grade results show U.S. students above the international average in both mathematics and science. In fact, U.S. fourth-graders are second only to Korea among their international peers in science. These results potentially point to a "slippage" problem in the U.S. that begins by the upper elementary grades. A more complete picture will hopefully emerge as data for grade twelve are released later in 1997 and early 1998, along with state-by-state comparisons of student performance in the U.S. and other participating countries. As it stands thus far, however, the main message for American education seems to be the need to teach and learn differently in the classroom. Class time should be used more productively for instruction, with lessons that are more coherent/focused and probe more deeply into the mathematics being studied, accompanied by classroom questioning strategies that elevate expectations for student learning.

Lesson content and structure matter. What teachers do in the classroom – and how they do it – matters.

Importance of Deductive Reasoning and Mathematical Proof

Structured lessons that are coherent, delve deeply into mathematics, and assess the degree of associated student understanding are critical to the improvement of classroom instruction and student proficiency. Reasoning, justification, and proof -- in an age-appropriate context -- are central to this mathematical coherence and depth. Unfortunately, what one observes all too often in small-group, student investigations in mathematics is simply the examination of cases or examples to solve a problem in a numerical context and perhaps to discover a numerical pattern. While these may be necessary steps for student understanding of mathematics, they are -- in the end -- largely inductive tasks. For student work to stop here is to miss the essence of mathematics.

Induction is important, for it is through careful verification of many examples that patterns are observed and conjectures are formulated; but, this work must lead to age-appropriate mathematical abstraction and to generalizations that are ultimately proved -- or disproved -- for some universe. The centrality of proof and deductive reasoning is what distinguishes mathematics from other disciplines. If we are genuinely to assess student understanding of mathematics, then we must place appropriate emphasis on proof and deductive reasoning in the classroom assessment of student work. A number of questions merit reflection:

- What constitutes "proof" at different grade levels, and what assessment rubrics are appropriate?

- Once students achieve a measure of mathematical maturity and are capable of providing arguments/proofs grounded in some degree of rigor, does that become the standard against which all their work should be assessed? That is, is it important to prove every result? If not, what factors inform a teacher's decision to have students prove one result and not another? When a teacher consistently fails to convey expectations that students prove or justify their work, is there a relationship to the teacher's own mathematical and pedagogical preparation?
- How does the teacher provide a classroom environment that conveys to students the centrality of justification/proof and deduction, so that it becomes a natural part of what they expect of themselves as they study and learn mathematics?

The TIMSS video tapes of middle grade classrooms in the U. S., Japan, and Germany provide interesting findings that may inform responses to some of these questions. The tapes have been studied to determine whether each concept presented to the students was only stated or whether it was also developed (U.S. Department of Education, 1996, p. 43). James Stigler, the lead investigator for the TIMSS video tapes project, provided an example to illustrate the notion of a *stated* concept (Bracey, 1997, p. 657):

> *Stated Example.* When a^m is divided by a^n [one may assume that a is a non-zero real number and that m, n are positive integers], the result is a^{m-n}. Once this property has been stated, the teacher asks the students to complete some exercises that apply the concept.

In contrast, the concept would be *developed* as follows (p. 657):

> *Developed Example.* The teacher provides an explanation of why the division of a^m by a^n results in a^{m-n} [for nonzero a] by writing and discussing with the class the following derivation:
>
> $$\frac{a^5}{a^3} = \frac{a \cdot a \cdot a \cdot a \cdot a}{a \cdot a \cdot a} = a^2$$

When *stated* versus *developed* concepts were studied in the videotapes, these average percentages were found (U.S. Department of

Education, 1996, p. 44): in Germany, *stated* 23% and *developed* 77%; in Japan, *stated* 17% and *developed* 83%; in the U.S., *stated* 78% and *developed* 22%. What emerges is a picture of classroom instruction in the middle grades with significant emphasis on explanation and justification in Japan and Germany and very little such emphasis in the U.S.

Classroom Questioning Strategies

One way a teacher can shape a classroom environment that emphasizes and values justification/proof and deductive reasoning is by the skillful use of questioning strategies that both set an appropriate expectation and provide a concrete means for assessing student understanding.   Questions such as the following, when posed to students after they have described a pattern or produced some result, are potentially useful:

- Does your pattern or result always hold? Can you think of an example where it doesn't? If "yes," can you think of another example? This may help the student decide whether there is only a single counterexample or a potentially "large" set of counterexamples.
- How will you generalize your pattern or result? For what universe?
- How do you justify the truth of the generalization on this universe?
- Can you define a larger universe on which your generalization holds?

Meaningful classroom discourse that weaves such questioning strategies into the discussion requires that the teacher work with students to ensure their understanding of such notions as *true* and *false*, the quantifiers *some* and *all*, *conditional* (or *if-then*) statements, and *counterexample*. By the time students are in high school, their understandings should be expanded to include many of the notions of elementary formal logic, including *converse, inverse, contrapositive,* and *equivalent statements*, as well as the commonly used methods of proof.

Assessment of Student Understanding

As students progress from describing patterns to formulating generalizations/conjectures to proving/disproving these conjectures, the assessment of their understanding requires careful reflection and an analysis grounded in the language and processes of mathematics as a discipline. Questions such as the following are integral to such assessment:

- To what degree is the generalization or conjecture posed by the student actually an abstraction of the pattern observed? In other words, to what degree does the conjecture accurately generalize the pattern recorded earlier by the student?
- To what extent does the universe for which the generalization/conjecture is stated take into account all examples previously investigated by the student? In other words, has the universe been defined to include the examples for which the generalization holds and exclude those for which it fails?
- In testing the conjecture, how varied and representative have been the examples selected by the student?
- Did the student provide a valid argument to justify or prove the conjecture?
- To what extent did the student's argument use relevant axioms or previously proved results/theorems? To set a classroom context that promotes coherent, in-depth study of mathematics, most lessons should be designed to extend or deepen the mathematics previously studied.
- To what extent does the student show an age-appropriate mastery of the language of mathematics, including terminology, notation, correct use of quantifiers, etc.?

These questions can provide a reasonable base for day-to-day, informal classroom assessments and, when coupled with appropriate grading rubrics, for more formal assessments of individual students.

A Classroom Example

Suppose a ninth-grade class has been studying a topic in discrete mathematics that includes finite, simple graphs (no loops nor multiple edges) and knows that, in general, a graph consists of a set of vertices and a set of edges connecting various vertices. The term *degree of vertex v* has been defined in graph *G*, as has the *total degree* (the sum of the degrees of all the vertices) of *G*. The class has been asked to investigate the relationship between the total degree of a graph and the number of edges in the graph. After exploring some examples, most of the student groups state -- in some form -- this conjecture: the total degree of a graph is twice the number of edges in the graph.

Group 1 presents its work as part of a formal assessment. The group had considered a half dozen or so pictorial examples of graphs, all connected (Figure 1).

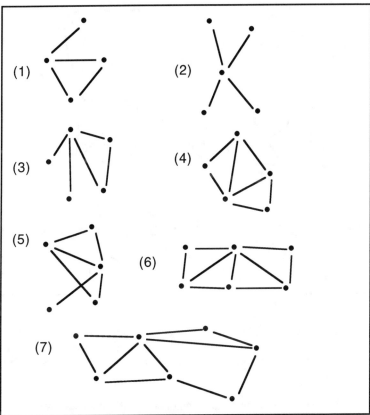

Figure 1. Examples Submitted by Group 1 Students

They then stated this conjecture: the total degree of any graph is twice the number of edges in the graph. The students proved their conjecture by noting that if e is an edge of graph G and e has endpoints v and w, it contributes 1 to the degree of v and 1 to the degree of w. So, every edge e contributes 2 to the total degree of G. Therefore, the total degree of any graph is twice the number of edges.

In Group 2, the students considered a half dozen or so pictorial examples that included both connected and disconnected graphs, as in

Figure 2. One graph even had isolated vertices. They stated the same conjecture as Group 1 and argued its truth similarly.

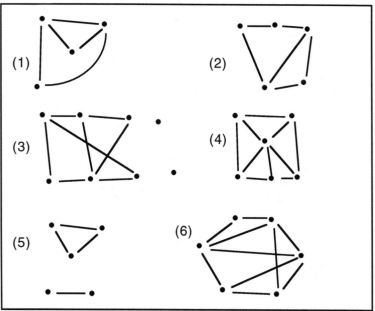

Figure 2.   Examples Submitted by Group 2 Students.

Group 3 considered several pictorial examples of graphs (see Figure 3) and stated the following conjecture: the total degree of a graph is always an even number. The students were unable to prove their conjecture.

There are several important components to the students' work:

1. The breadth of examples considered.
2. The conjecture, including the universe on which it is stated and the degree to which the conjecture is a faithful generalization of the examples.
3. The justification or proof of the conjecture, including use of relevant axioms or previously proved results/theorems.
4. The communication of the mathematics, including terminology, notation, and correct use of quantifiers.

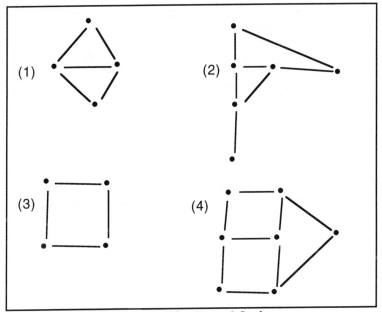

Figure 3. Examples Submitted by Group 3 Students.

Decisions, informed by the values of the discipline and a knowledge of the goals or purposes of the lesson, must be made about the relative importance of each of these four components. Absent special factors related to the goals/purposes of the lesson itself, one should expect the second and third components to command greater weight than the first and fourth.

We might use a five point scoring rubric, where *0* represents an inability to begin, *1* represents an appropriate beginning but with subsequent major flaws (i.e., the work collapses) or inability to complete the work, *2* represents completion of the work but with minor flaws (e.g., omission of a trivial or "nonfatal" case in a proof involving case exhaustion, or an imprecise explanation), *3* represents a correct completion of the work, and *4* represents an exemplary completion of the work. Scores of *0* and *1* denote responses that may be categorized as unsatisfactory; *2*, as satisfactory; and *3* and *4,* as competent.

In this context, the work of Group 1 corresponds to a score of *2* on the first and second components, a *4* on the third and fourth components, and an overall score of *3*. The students' examples (all connected graphs) were not sufficiently representative and varied to support the conjecture they stated. The work of Group 2 corresponds to a *4* on all individual components, as well as an overall *4*. The work of Group 3 corresponds to an overall score of *1*. The students made an appropriate beginning with several examples and stated a conjecture. As with the first group of students, however, the examples presented by Group 3 were not sufficiently representative and varied to support the general conjecture. In addition, the students were unable to complete their work.

Some Caveats

First, there is some sense among practicing mathematicians that a prevailing philosophy in the mathematics education community today "portrays mathematics ... as essentially fallible -- it replaces the notions of truth and objectivity with a kind of social acceptance," a view of the discipline not generally shared by mathematicians (Selden & Selden, 1997). Asking the class such questions as "how many of you think Marty has proved such-and-such-a-result?" -- followed by a showing of raised hands -- potentially sends a message to novice learners that sound mathematics is socially defined, which it is not. It is perhaps better to ask "has Marty proved such-and-such-a-result?" or "is Marty's argument convincing?", followed by "what leads you to that conclusion?" or "are there any places where Marty's argument breaks down?". Mathematics is a dynamic discipline that grows and expands to include new results, but such notions as mathematical validity and consistency are the bedrock of the discipline.

Second, teacher expectation about mathematical proof and justification needs to be tempered with an understanding of what is age-appropriate. As students progress through high school, however, they should be exposed to increased rigor and a variety of different methods of mathematical proof, including proof by contradiction and proof by contrapositive.

Acknowledgment

The opinions expressed here are those of the author and do not necessarily reflect those of the National Science Foundation.

References

Beaton, A. E., Mullis, I. V. S., Martin, M. O., Gonzalez, E. J., Kelly, D. L, & Smith, T. A. (1996). *Mathematics achievement in the middle school years: IEA's Third International Mathematics and Science Study (TIMSS)*. Chestnut Hill, MA: Boston College.

Bracey, G. W. (1997). More on TIMSS. *Phi Delta Kappan*, 78(8), 656-657.

Mullis, I. V. S., Martin, M. O., Beaton, A. E., Gonzalez, E. J., Kelly, D. L., & Smith, T. A. (1997). *Mathematics achievement in the primary school years: IEA's Third International Mathematics and Science Study (TIMSS)*. Chestnut Hill, MA: Boston College.

Selden, J., & Selden, A. (1997). Mathematicians' views of mathematics, (preliminary report). *Abstracts of papers presented to the American Mathematical Society, 18*(1), 173.

Schmidt, W. H., McKnight, C. C., & Raizen, S. A. (1996). *Splintered Vision: An Investigation of U.S. science and mathematics education: Executive summary*. East Lansing, MI: Michigan State University, U.S. National Research Center for the Third International Mathematics and Science Study.

Stigler, J., & Manaster, A. (1997, January). *Third International Mathematics and Science Study (TIMSS)*. Presentation at Joint Mathematics Meetings, San Diego, CA.

U.S. Department of Education, National Center for Education Statistics. (1996). *Pursuing Excellence (*NCES 97-198). Washington, DC: U.S. Government Printing Office.

Open-ended Tasks: A Key to Mathematics Assessment

Chuck Thompson
University of Louisville

Open-ended tasks are tasks which have a variety of successful responses. This can occur in at least two ways. The task may pose a problem that can be solved in a variety of ways, or there may be a number of possible correct answers. The vehicle problem (Figure 1) is an example of a problem that may be solved in several different ways, and it has a number of possible correct answers. The solution methods range from using counters as wheels to model the situation, to making a table in which the various vehicles are listed at the tops of columns and the numbers of each of those vehicles used are placed in the corresponding columns.

Other desirable characteristics of open-ended tasks are that they have a low entry level and a high ceiling. A low entry level enables all or almost all students to become engaged in the problem and demonstrate mathematical ability. A high ceiling encourages able students to demonstrate as much mathematical power as they can. For primary level students, the vehicle problem (Figure 1) has been found to satisfy these criteria as well as the defining criteria described in the previous paragraph.

In the parking lot there is a group of vehicles having a total of 18 wheels. Possible vehicles are:

Bicycle	Car:	Bus:
2 wheels each	4 wheels each	6 wheels each

What group of vehicles could be in the parking lot? Find as many possibilities as you can. Use words, pictures, and numbers to show how you got your answers any how you know that they are correct.

Figure 1. The Vehicle Problem

Recognizing the Value of Open-ended Tasks

One significant value in assessing with open-ended tasks is that the focus is placed on what students can do rather than on what they cannot do. The low entry level aspect of open-ended tasks enables students to demonstrate the mathematics that they know; the high ceiling characteristic enables students to demonstrate all that they know.

Another important reason for using open-ended tasks is that solution methods are valued as much as answers. When a variety of solution methods is possible, the student does not have to recall or use any particular one. In this way, students can respond to the task in ways that they feel most secure and comfortable.

A third reason for using open-ended tasks is that reasoning and mathematical communication can readily be assessed. This is accomplished by asking students to explain their reasoning as they solve the problem, or after they have finished. This is done by attaching a statement such as the last one in the vehicle problem: "Use words, pictures, and numbers (or symbols) to show how you solved the problem and how you know that you are right." Note that this requires students to do more than simply recall and restate their steps in solving the problem. It requires them to analyze their solution and justify its correctness, a process that involves higher level thinking and mathematical reasoning.

Creating Open-ended Tasks

Teachers can create open-ended tasks by modifying existing tasks. The following list provides some strategies for doing this:

- explain their work to a friend (or younger person)
- draw a diagram to show how they solved the problem
- use manipulatives to solve the problem
- solve the problem in more than one way
- explain how they know that their answer is correct
- describe a different problem that is like the given problem in a particular way
- tell how the given problem could be used in real life
- make up a word problem to match a given computation problem
- make up a similar problem that is harder/easier
- draw a diagram to show their answers to a given problem
- change the conditions given in a specific problem.

In Figure 2 there are two conventional tasks which have been modified to make them open-ended. Notice that the level of thinking that is required to solve the modified tasks is much higher than in the conventional tasks.

Conventional Task	Modified (Open-Ended) Task
Solve: 8 x 14 = _____	Solve 8 x 14 = _____ in three different ways by using other multiplication facts. Show how you get the answer each way and how you know it is correct.
Armando's test scores are 76, 66, 92, and 82. What is his average score?	Armando's test scores are 76, 66, 92, and 82. What must he score on his next test to have an average test score of 80? Explain how you know you are right.

Figure 2. Modified Tasks

Scoring Responses to Open-ended Tasks

A good way to score responses to open-ended tasks is to use a scoring guide which addresses the important features of student performance that are assessed through open-ended tasks. These features of student performance are mathematical knowledge, and, as mentioned previously, reasoning, and communication ability. The following general scoring guide addresses these features (Figure 3).

Level 4	The student completes all parts of the task completely and communicates his or her solution effectively. There is evidence of in-depth understanding and logical reasoning.
Level 3	The student completes most important parts of the task and communicates adequately. There are minor flaws in either the solution of the problem, reasoning used, or the communication of the solution to the problem. Some minor aspect of the task may not be addressed.
Level 2	The student completes some important parts of the task and communicates them adequately. There is evidence of significant gaps in the student's mathematical knowledge, logical reasoning, or ability to communicate that knowledge.
Level 1	There is some understanding of the mathematics involved in the task, but it is minimal.
Level 0	There is no evidence of understanding of the mathematics involved in the task.

Figure 3. A General Scoring Guide

The scoring guide (Figure 3) exhibits the characteristics of good scoring guides. These characteristics are the following:

- A good scoring guide describes the desired qualities of student performance,
- A good scoring guide specifies 3-6 performance levels,
- A good scoring guide is clearly stated so that students know what is expected of them, and
- A good scoring guide explicitly distinguishes one performance level from another.

The scoring guide shown in Figure 3 is general. It can be used to write scoring guides which are specific to problems actually given to students for assessment purposes.

The scoring guide in Figure 4 corresponds to the vehicle problem presented earlier (Figure 1). It is written for students in grade 4, and could be modified for students in other grade levels by changing the descriptors regarding the number of solutions found, the quality of the reasoning used, and the quality of communication included.

Score	Description
4	Nine or more solutions (out of 12 possible) found. Evidence of a systematic approach to solving the problem. Solution method is communicated effectively.
3	No evidence of a systematic approach to solving the problem, but at least six solutions are found. Communication of solution method is satisfactory.
2	Six or more solutions are found, but without explanation, OR four or more solutions are found with satisfactory communication of solution method.
1	Some evidence of combining multiples of 2, 4, and 6 in attempting to get a sum of 18.
0	No evidence of understanding.

Figure 4. Scoring Guide for the Vehicle Problem

Identifying Characteristics of Good Open-ended Tasks

The very best open-ended tasks have additional characteristics to the those described in the first section of this paper: (a) multiple solution methods and/or multiple answers, (b) low entry level, and (c) high ceiling. But, the best open-ended tasks are also (d) engaging to students. The student is motivated to respond to the task because it is interesting or intriguing. The best tasks also (e) have many possible connections -- to real life, to other subjects, or to other mathematics. For example, the vehicle problem is potentially interesting to students because it is a problem that could reasonably exist near their own school. The best tasks also (f) involve mathematics that is central to the curriculum.

The open-ended task presented here (Figure 5) also is a very good one because it involves the characteristics just described. The vehicle task involves addition of whole numbers and multiples of whole numbers, certainly content that is an important part of the mathematics curriculum. Furthermore, the thinking that is required to find (almost) all of the solutions involves the type of reasoning that all students should be expected to demonstrate.

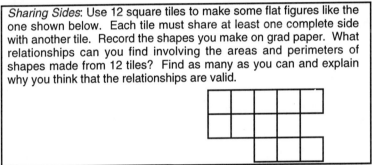

Sharing Sides: Use 12 square tiles to make some flat figures like the one shown below. Each tile must share at least one complete side with another tile. Record the shapes you make on grad paper. What relationships can you find involving the areas and perimeters of shapes made from 12 tiles? Find as many as you can and explain why you think that the relationships are valid.

Figure 5. An Open-ended Task

Matching Instruction with Open-ended Assessment

In order to prepare students for performing successfully on open-ended assessment tasks, teachers will want to use open-ended tasks on a regular basis during instruction. One way to do this is by modifying conventional tasks, as was described in the first section of this paper. Another way is to find open-ended tasks in textbooks, in resource books, or in literature from professional organizations, such as the National Council of Teachers of Mathematics. If students learn through the use of open-ended tasks, then the assessment will match the instruction that is used.

Equally as important as using open-ended tasks on a regular basis is the use of open-ended verbal questions during class discussions. Questions which ask students to justify their thinking and to explain their reasoning will help students become more capable of higher level thinking. Likewise, asking students to share alternative methods of solving problems will help them understand that no single method is preferred over others, and that they can choose a method that is comfortable to them. Furthermore, by involving students in higher level thinking, teachers can help students enjoy and appreciate mathematics as a subject that is much more that the rote memorization of rules and formulas.

Conclusion

In this paper some practical considerations for using open-ended tasks have been described. Guidelines have been given for creating such tasks and for scoring student responses to those tasks. Characteristics of good open-ended tasks have been shared. And, reasons were given for assessing through the use of open-ended tasks. But, perhaps the main reason that open-ended tasks should be used for assessment is that they assess what mathematics educators believe is most important for students to be able to do to solve problems, to reason, and to communicate mathematically.

Clear Learning Targets

Jan Williamson
North Carolina Department of Public Instruction

Good instructional planning begins with the teacher's establishing clear learning targets (what students should be able to understand and to do) and then constructing precise expectations or standards (how well students need to be able to perform). The North Carolina Mathematics Curriculum has seven strands: numeration, geometry, patterns, measurement, problem solving, data, and computation. In any one of these mathematics strands, the learning target would be *what* the student would be expected to understand or to do. In problem solving, for example, an eighth-grade student would be expected to formulate new problems from existing problem situations. The standard should be *how well* the student is able to do that task; the measure for the standard could be dimensions such as independence, accuracy, or consistency. Thus, a student who *independently, accurately, and consistently* formulated new problems from existing problem situations would be achieving on a high level, while a student who needed *frequent assistance* or guidance and who *rarely* completed the task accurately would be scoring on a low level.

After formulating and clarifying the targets and expectations, the teacher can then work backwards, asking the essential questions: How can I design assessments that will demonstrate if these learning targets are being met? What instructional activities should I plan so that all students can reach these targets?

While establishing clear learning targets sounds simple, it is remarkably complex, partially because of how difficult it is to differentiate clearly what the learning target is. Often the target gets confused with the instructional activity which the student is asked to engage in. For example, using a graphic organizer to solve a word problem is an activity, not a learning target. The target is the ability to read and solve word problems, a subgoal is to identify necessary information and conceptualize how that information is related, and the activity is designing a graphic organizer.

A learning target, then, should be a point of evaluation that indicates student achievement. It should include only enough specifics for clarity and should require all students to demonstrate high, but appropriate, outcomes. In addition to the preceding definition, the following criteria can be useful in establishing learning targets:

- Good learning targets must be *complete* in that they should include the skills, the conceptual understandings, and the performances that students need. They should not leave out any essential performances or understandings that the student will need to move to his/her next level of accomplishment.

- In addition, they must be *essential*. It is often difficult to make sure that the target does not have gaps and at the same time does not include non-essentials. For example, a learning target that focuses on the student's ability to solve a problem, without including an assessment of the student's conceptual understanding, is not complete because it does not determine whether the student is thinking like a problem-solver or merely following an algorithm. Correspondingly, a target that asks a student to follow a particular sequence of steps in problem-solving includes non-essentials.

- Good learning targets should be *capable of being assessed* in the classroom (observed, measured, etc.); the assessment should demonstrate clearly whether or not students understand a concept or can perform a task. As our understanding of classroom assessment expands to include teacher observations, self-observations, and peer-evaluations, we find more and more useful tools with which to observe and measure important, but traditionally ambiguous targets, such as perseverance.

- Good learning targets should be *conceptualized at an application level* that subsumes subskills and details. Isolated, discrete skills, and pieces of information should not be assessed out of context, but rather in context at the application level. Once again, the target should reflect the important conceptual understandings and performances; the skills and factual details are important as they are subsumed in the target.

- Clear targets should *spiral* through the curriculum; they should not be discrete objectives which are applicable only to one grade level. For this reason, students should see connections between the targets of different grade levels. However, the level of sophistication and complexity will increase as the student advances through grades. The North Carolina mathematics curriculum, with its seven strands that spiral through grades K-12, provides an excellent example.

High levels of student autonomy, ownership, and input are essential for significant learning; in fact, they are hallmarks of excellent education and quality assessment. Michael Scriven, for example, delineates the following levels of student involvement and ownership in the assessment process, beginning with a low level and moving to increasingly higher levels:

- take the test and receive the grade
- be invited to offer the teacher comments on how to improve the test
- suggest possible assessment exercises
- actually develop assessment exercises
- assist the teacher in devising scoring criteria
- apply scoring criteria to the evaluation of their own performance
- come to understand how the assessment and evaluation processes affect their own academic success
- come to see how their own self assessment relates to the teacher's assessment and to their own academic success. (Scriven, as cited in Stiggins, 1997, p. 18)

Scriven's list illustrates how important student ownership can be for student insight and understanding, and thus, achievement. However, note that even the highest levels of student ownership do not include the student's creating or defining the learning targets.

Rather, the curriculum and the teacher, not the student, should establish the learning targets. The student should understand these targets clearly and precisely. She or he should then have great responsibility in self-assessment; in giving the teacher feedback about the additional instruction he or she needs next; and in choosing materials, activities, projects, and demonstrations of learning. The learning targets, however, should be pre-established and thus consistent and stable, as well as clear. Students generally do not have the expertise or experience to establish learning targets for themselves;

however, they can almost always hit any target which they can see clearly and accurately.

Reference

Stiggins, R. (1997). *Student-centered classroom assessment.* Upper Saddle River, NJ: Merrill.

Classroom Assessment: View from a Second-Grade Classroom

Lisa S. Williamson
Pinnacle, NC, Elementary School

Classroom assessment has many different definitions according to the classroom teacher implementing the assessment process. In my classroom it is lots of different ideas that mesh together to create an understanding of the child's learning and where we need to go next. First and foremost, classroom assessment must be used for planning further instruction. Often we hear teachers discussing the need to "cover" information with their children. Unless the children are ready for the educational journey we are taking them on, they are not going to be able to construct meaning from the tasks. Therefore, I believe we must constantly assess our children as they travel with us on this journey.

Based on my experiences as a second-grade teacher I feel teachers need to collect data about their children in many different ways and at many different times. Data should be collected daily in a formal or informal way. It is hard to eat lunch with the children without constantly observing and assessing the way they use math in their environment or the communication skills they exhibit with their friends. Good teachers use every minute of the day to build knowledge about the children that will help them to make decisions about the learning of that child. There are many forms of record keeping teachers can use to document student growth or lack of understanding. I collect information daily on a calendar grid. These are usually anecdotal notes concerning the tasks the children have been involved in and how they were able to complete the task. I write about children's reading and comprehension on index cards on a daily basis, and I consistently save work samples for portfolios. I pay special attention to those children who are having difficulty and those children who are showing an ability to go beyond the expectations we have set.

I also use paper-and-pencil tasks on a regular basis to help me see how the children are thinking. I often score these tasks using a rubric and share this information with the children to help them become better self-assessors. Our report cards are also based on a rubric system, so parents and children are very familiar with the scoring and expected outcomes. Pencil and paper tasks give the children opportunities to record their own answers and help to prepare them for more formal assessments as they move through the grades. Many of these paper-and-pencil tasks are also saved for the children's portfolios. These documents are then explored by the children at various times during the year so they have an opportunity to see their progress, or lack of progress, on various mathematical topics as well as other subject areas.

I strongly believe in the use of individual interviews in making decisions about student growth. I have informal individual interviews with my children during the school day and always spend a large block of time with my children individually at the end of each grading period during a more formal interview. At this time I use performance tasks with the children and allow them to demonstrate their knowledge in completing these tasks. This information, along with other anecdotal evidences and written tasks the children have completed, are then synthesized and a decision is made as to where to mark the child on a Mathematics Matrix. The matrix I use is the one available from the North Carolina Department of Public Instruction.

In my classroom, I have implemented a concept we refer to as "Round About Math." During Round About Math the children are involved in several center type activities in which the objectives that have been taught the last nine weeks are made into games or assessment activities for the children to complete. As the children move around to the various activities, I am able to call the children individually, or two or three at a time, and spend quality time questioning them and asking them to show me certain behaviors I need to see exhibited before making a decision as to where to mark them on the matrix. I see this time spent with the children as very valuable because I am able to determine exactly what they know and what I need to do with them. These individual interviews are the highest quality assessment of all the information I have gathered. I spend time with the children allowing them to see where they are on the matrix so they will know what the expectations are for the next grading period. If there are things they were not proficient in the last grading period, we go back and reassess those things and celebrate our progress together. If there are weaknesses showing up now, we discuss the need to spend more time on those areas and what we can do together to improve. We also are clear about what will be communicated to parents, and the children know what will be marked on their report card. Because of this, children are often coming to me and saying things such as "Now I know how to do 10 more and 10 less, my dad helped me and we did it with dimes." If the children take an interest in making progress, very often parents will spend the extra time to make the gains they need.

Appropriate classroom assessment is the key to good planning and good instruction. Professional development with long term follow-up is needed to help teachers learn how to get behind the eyes of their children and know how to help them make strides in their understanding. Teachers often feel they do not have time to assess children individually because they should be "teaching." We must help those teachers to see the benefits of knowing and understanding their children in order to "teach" them. Professional development that would give teachers the opportunity to share success stories and work together would create a positive attitude toward classroom assessment. But where is the time?

I believe that the expectations we set for our children are the framework with which we can build great things. We must make sure that all children are nurtured and challenged at the same time. This is a very difficult, but very necessary, task. Without being both nurtured and challenged, our children will not make appropriate progress each year. The expectations in my classroom are set by the Standard Course of Study; however, I want my children to exceed the Standard Course of Study whenever possible. Of course there are those children who can immediately exceed and those children who will need to be nurtured throughout the year. I see my job as taking everyone as far as I possibly can. My knowledge of clear targets for all children and my knowledge of the expectations of the Standard Course of Study enable me to know what to look for and what to build on as we work throughout the year. All children can learn, and with good assessment practices all children can be made aware of their learning and the advancements they make over time.

Assessing Conceptual Understanding

Judith S. Zawojewski
National-Louis University and University of Pittsburgh

Edward A. Silver
University of Pittsburgh

Any good mathematics program would have as one of its goals enhancing student understanding of important mathematical concepts. When teaching concepts, teachers routinely embed a given mathematical idea in a variety of contexts, situations and performance circumstances. However, *assessment* of a given conceptual understanding is oftentimes based on the results of one item on a test. Yet, teachers report that the very same students who answer the test item correctly are often unable to recognize the concept in a new situation or cannot apply the concept in a problem-solving setting. This experience leads to a series of questions for classroom practice: What does it mean to *know* a concept? How can teachers *know* that a student knows a concept?

As Hiebert and Carpenter (1992) note, the assessment of understanding is a highly inferential task relying on indirect inference from performance. For this reason (i.e., to reduce the likelihood of error due to over-inference) and also because of the inherent complexity of mathematical understanding (cf., Hiebert, 1988; Sierpinska, 1994; Skemp, 1987), "understanding usually cannot be inferred from a single response on a single task.... A variety of tasks, then, are [sic] needed to generate a profile of behavioral evidence" (p. 89). A recognition of the need for multiple measures of understanding is at the heart of assessing a concept. If the multiple measures can be systematically generated to represent different aspects of "knowing," documentation of student performance could potentially generate profiles of understanding. Toward that end it is useful to consider two hallmarks of genuine mathematical understanding: the robustness of a student's understanding of a concept in different settings (e.g. varying task features and/or performance conditions), and the connection of the targeted concept to other mathematical ideas.

Probing Students' Understanding: An Example

We begin by examining an individual assessment task and specifying the mathematical understanding it is intended to assess. When considering how the task may be used to measure the identified concept, we need to avoid the danger of over generalizing what it is that a student knows. Using the recommendation of Hiebert and Carpenter, we can consider generating a collection of other tasks meant to assess the same mathematical idea, perhaps providing us with an enhanced view of student understanding.

To illustrate, consider a released item from the 1990 National Assessment of Educational Progress (Figure 1). Linda's Boxes (Figure 1) could be said to assess a mathematical idea that appears in multiple places across the elementary and middle school mathematics curriculum: that there is a covarying relationship between the size of a unit and the number of units needed to represent a given amount.

Robustness of Understanding

Research has established the fragility of understanding with respect to varying task features. For example, in mathematics and science assessment, Baxter, Shavelson, Herman, Brown, and Valadez (1993) reported significant variation in performance for tasks that differed only slightly in their settings. Inconsistencies in performance on purely symbolic tasks as compared with story problems have been noted for the past half century, and a contemporary version of this phenomenon is documented in research that probes the relation between school learning and performance in non-school settings (e.g., Nunes, Schliemann, & Carraher, 1993). With respect to varying the mathematical entity, a number of researchers (e.g., Greer, 1992) have documented the limits in children's understanding of multiplication and division by examining their responses to tasks in which the entity being operated upon is varied from whole numbers (for which "multiplication makes bigger and division makes smaller") to fractions or decimals (for which this apparent generalization no longer holds). The fragility of understanding in the face of variation in the size of numbers is well-documented in research on errors in using algorithmic procedures (e.g., Brown & Van Lehn, 1982) as well as in research aimed at more conceptual targets (e.g., Brownell, 1935). Finally, assessing a mathematical concept directly is quite different than assessing it indirectly. For example, asking students to identify the right triangle in a set of triangles assesses the concept of right triangle directly. On the other hand, an indirect assessment of the concept occurs in a task which does not explicitly ask for a right triangle to be identified, but requires that students do so in order to solve the problem given.

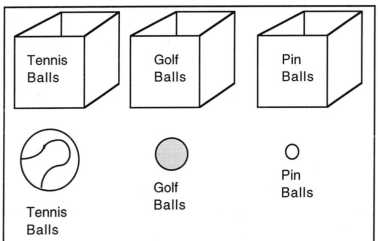

Linda has three large boxes all the same size and three different kinds of balls as shown above. If she fills each box with the kind of balls shown, which box will have the fewest balls in it?

A. The box with the tennis balls
B. The box with the golf balls
C. The box with the pin balls
D. You can't tell

Source of Question
1990 NAEP Mathematics Assessment - Grade 4

(NOTE: The task was slightly revised, changing "rubber balls" - which implies no specific size - to "pin balls." The labeling was also altered to make it more readable and clear.

Figure 1. Linda's Boxes

The robustness of students' understanding of a concept across varying task features can be inferred when they use the mathematical idea in a variety of context or settings, when there are variations in the size of the objects involved, or when the concept is embedded in different mathematical entities. (With regard to contexts, while it would seem that a concept is more robust when it can be adapted for use in unfamiliar contexts, defining "familiar" and "unfamiliar" for a population is problematic.) An illustration of task feature variation can be found in Figure 2.

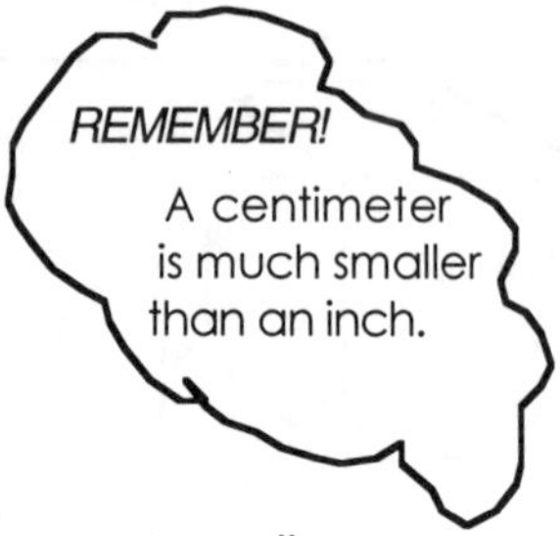

REMEMBER!
A centimeter is much smaller than an inch.

Two students measured the length of the same pencil.

Adam said, "The measure of the pencil is *10*."
Leah said, "The measure of the pencil is *4*."
The teacher said, "You both may be right!"

One of the students used *centimeters* and one of the students used *inches*.

1. Write *centimeters* or *inches* in the blanks:

    Adam should have said: 10   _____

    Leah should have said: 4   _____

2. Explain how you know which should be centimeters and which should be inches.

Figure 2. Length of a Pencil

The assessment task in Figure 2 varies from the one in Figure 1 because the mathematical entity has changed from volume to length, yet both assess the targeted concept: that the size of the unit covaries with the number of units needed to represent a given amount. The Length of a Pencil task addresses this concept with respect to the attribute of length instead of volume by asking students to explain their selection of a unit of length. One can imagine assessing this concept also across other aspects of measurement, such as time or area.

Another way to examine the robustness of a student's understanding is to assess it under different performance conditions. Much of the contemporary interest in performance assessment is based on considerable evidence that performance can be influenced by the conditions of the assessment. For example, Saxe (1991) who studied young Brazilian candy sellers, found that they performed computational activity nearly flawlessly in everyday activities but made errors when presented with structurally equivalent paper-and-pencil tasks. Differences in the availability of amplifiers of student performance, such as human resources (e.g. working with partners) and physical resources (e.g. calculators, manipulatives), can also lead to differential performance. Evidence drawn from diverse sources, including research conducted within a sociocultural paradigm and data available from student performance on the National Assessment of Educational Progress (Kouba & Swafford, 1993), indicates variability in student performance on tasks when calculators are or are not utilized, when human or intellectual resources may or may not be consulted, and when physical models are or are not available. Note that the Length of a Pencil (Figure 2) varies slightly from Linda's Boxes (Figure 1), with respect to performance conditions, because it requires students to construct a response rather than select one from a list of choices. Performance conditions can also vary more dramatically, as is illustrated in the task in Figure 3. Like the first two tasks, the Pan Balance Interview (Figure 3) assesses the given targeted concept. Unlike the first two tasks, which provide no outside resources to the student, the task in Figure 3 provides students with physical resources in the pan balance and manipulatives along with prompting from an interviewer. Not only would one expect different performance levels on these three tasks, but different types of information could be inferred from each, although the target concept remains constant.

Connections to Related Mathematics

It is widely recognized that mathematical ideas do not exist in isolation from each other, and a recognition of and facility with the interrelationships among sets of mathematical ideas is a basic feature of genuine mathematical understanding. One type of connection occurs when an assessment task requires that a variety of mathematical ideas be used in the solution process. In these cases, the reason different mathematical concepts are embedded side-by-side is largely due to the problem context. For example, the task in Figure 4, requires not only the concept of the covarying relationship between the size of a unit and the number of units needed to represent an amount, but also real world experience with measures of 12-year-old students, and an understanding of the variety of units that may be used in this situation. So, one might say that the targeted concept is embedded along side a number of other concepts in this particular assessment task.

Materials: a pan balance (simple fulcrum model)
                   15 one-gram weights
                   1 five-gram weight

Interview Prompts

1. Each of these is a one-gram weight. (Show student the one-gram weight). *What is the mass of this weight?* (Show student the five-gram weight. If necessary prompt the student to find the weight of the five-gram weight using the pan balance and the available weights.)

2. Alice put *15* one-gram weights in the right side. She wants to put enough five-gram weights into the right hand side in order to balance the pans. Will Alice use *more than 15, less than 15, or exactly 15 five-gram weights* on the right side? (Notice that there are not enough five-gram weights for students to act this out, forcing the student to reveal his or her reasoning.)

3. Explain how you know. (As the student works, additional prompts may be necessary to help the student articulate his or her reasoning.)

Figure 3.   Pan Balance Interview

Interview:

A teacher posed the following problem to her class:

    I saw a 12-year-old student yesterday whose age was 144!
    The student was 144 what?

Solve this problem.
Explain your thinking aloud as you work towards a solution.

Figure 4.   144 What?

A second type of mathematical connection might be described as one in which the targeted concept, as it is used in the original task, needs to be modified to solve a different task. To illustrate, consider again the tasks in Figures 1 and 4. In Figure 1, students were given a fixed amount to represent (the identical boxes) and the relative sizes of the unit, and they were asked about the number of units needed. The task in Figure 4, on the other hand, provides the students with the number of units, and asks them to determine the size of the unit. Further, the amount being represented is not given directly; instead students must infer what amount is being represented by simultaneously considering attributes of a 12-year-old student and units of measure for each attribute. Thus, this task requires a modification of the use of target concept as it appeared in the first task. So, the mathematics connection is not to a *different* mathematical concept, but to a concept that is closely related to the one assessed in the first task.

A third type of connection would involve embedding the targeted concept in a different mathematical unit of study. For example, the Comparing Fractions Task in Figure 5 requires students to use the targeted concept with respect to the domain of rational numbers. In deciding the relative order of the fractions, students need to interpret the numerator as the number of units, and the denominator as the size of the unit. When the numerator and denominators are considered together (with respect to the same whole), then students can make an inference about the amount being represented by each fraction, and then compare them to determine the fraction that represents the least amount.

---

1. Which of the following fractions is the least? Circle it.

$$\frac{3}{4} \qquad\qquad \frac{3}{9} \qquad\qquad \frac{3}{5}$$

2. Explain how you know the fraction you circled is the least of the three listed.

Source of Task
Adapted from Behr, M. J., Wachsmuth, I., Post, T. R., Lesh, R. (1994). Order and equivalence of rational numbers: A clinical teaching experiment. *Journal for Research in Mathematics Education, 15,* 323-341.

Figure 5. Compare Fractions

So, perhaps we can consider three types of mathematical connections when generating task variations: when different concepts are side-by-side to solve a problem, when a concept needs be modified or adapted for use in a different problem, and when a concept resurfaces in different units of study. Granted, while all mathematical ideas can be linked to each other in the "grand scheme of things" and perhaps all mathematical ideas can be considered to exist on a continuum of connections, at this point in time it seems useful to make these distinction for the purpose of developing task variations and profiles of student learning.

Implications for Classroom Assessment

Assuming that a single assessment task is considered inadequate for assessing an important mathematical concept, one can envision a constellation, or collection of tasks, that assess related aspects of the idea. If a framework of dimensions for variation could be developed, it could serve not only to help in the generation of constellations of tasks organized around targeted concepts, but also to provide profiles of student understanding that would facilitate instructional decision-making. Further, the collection of assessment tasks could also serve as a forum of communication and discussion with students and their parents about the nature of their understanding. Not only would they be helped to conceive of conceptual learning as complex and rich, but they would also have a sense of the patterns of strength and weakness in a student's partial and growing knowledge of a concept. The development of this idea for assessment seems to have potential to serve multiple purposes for assessment, including making instructional decisions, students' self-assessing, reporting rich information about individual students' conceptual understanding, and communicating with interested audiences about important mathematical ideas.

Acknowledgment

This paper represents work conducted in the *B*enchmarks *o*f *S*tudent *Un*derstanding (BOSUN) Project, which is part of the National Center for Improving Student Learning and Achievement in Mathematics and Science (NCISLA). The work reported herein was supported under the Educational Research and Development Centers Program (PR/Award Number R305A6007), as administered by the Office of Educational Research and Improvement, U.S. Department of Education. However, the contents do not necessarily represent the position or policies of the National Institute on Student Achievement, Curriculum, and Assessment, the Office of Educational Research and Improvement, or the U. S. Department of Education.

References

Baxter, G. P., Shavelson, R. J., Herman, S. J., Brown, K. A., & Valadez, J. R. (1993). Mathematics performance assessment: Technical quality and diverse student impact. *Journal for Research in Mathematics Education, 24*, 190-216.

Brown, J. S., & Van Lehn, K. (1982). Toward a generative theory of "bugs." In T. P. Carpenter, J. M. Moser, & T. A. Romberg (Eds.), *Addition and subtraction: A cognitive perspective* (pp. 117-135). Hillsdale, NJ: Lawrence Erlbaum Associates.

Brownell, W. A. (1935). Psychological considerations in the learning and teaching of arithmetic. In W. D. Reeve (Ed.), *The teaching of arithmetic: Tenth yearbook of the National Council of Teachers of Mathematics* (pp. 1-31). New York: Teachers College, Columbia University.

Greer, B. (1992). Multiplication and division as models of situations. In D. A. Grouws (Ed.), *Handbook of research on mathematics teaching and learning* (pp. 276-295). New York: Macmillan.

Hiebert, J., & Carpenter, T. P. (1992). Learning and teaching with understanding. In D. A. Grouws (Ed.), *Handbook of research on mathematics teaching and learning* (pp. 65-97). New York: Macmillan.

Hiebert, J. (1988). *Conceptual and procedural knowledge: The case of mathematics.* Hillsdale, NJ: Lawrence Erlbaum Associates.

Kouba, V. L., & Swafford, J. O. (1989). Calculators. In M. M. Lindquist (Ed.), *Results from the fourth mathematics assessment of the National Assessment of Educational Progress* (pp. 94-105). Reston, VA: National Council of Teachers of Mathematics.

Nunes, T., Schliemann, A. D., & Carraher, D. W. (1993). *Street mathematics and school mathematics.* New York: Cambridge University Press.

Saxe (1991). *Culture and cognitive development: Studies in mathematical understanding.* Hillsdale, NJ: Lawrence Erlbaum Associates.

Sierpinska, A. (1994). *Understanding in mathematics.* London: Falmer Press.

Skemp, R. R. (1987). *The psychology of learning mathematics.* Hillsdale, NJ: Lawrence Erlbaum Associates.

Section 4

Appendices

Conference Participants

Harold Asturias, New Standards Project
Carne Barnett, WestEd
Mary Louise Bellamy, University of North Carolina Mathematics and
    Science Education Network
Sarah B. Berenson, North Carolin State University
Dee Brewer, North Carolina Department of Public Instruction
Diane Briars, Pittsburgh Public Schools
George W. Bright, University of North Carolina at Greensboro
William S. Bush, University of Kentucky
Gwendolyn Clay, Meredith College
Douglas H. Clements, State University of New York - Buffalo
Carolyn Cobb, North Carolina Department of Public Instruction
Lou Fabrizio, North Carolina Department of Public Instruction
Francis (Skip) Fennell, Western Maryland College
Susan N. Friel, University of North Carolina at Chapel Hill
Tery Gunter, Durham Public Schools
Sam Houston, North Carolina Department of Public Instruction
Audrey Jackson, Fenton, MO, Public Schools
Mazie Jenkins, Madison, WI, Public Schools
Henry Johnson, North Carolina Department of Public Instruction
Jeane M. Joyner, North Carolina Department of Public Instruction
Patricia Ann Kenney, University of Pittsburgh
Mike Kestner, North Carolina Department of Public Instruction
Mary Montgomery Lindquist, Collumbus State University
Toni Meyer, North Carolina Department of Public Instruction
Carol Wickham Midgett, Southport, NC, Public Schools
Vicki Moss, Wake County Schools and Randolph County Schools
Mari Muri, Connecticut Department of Education
Ruth E. Parker, Educational Consultant
Sid Rachlin, East Carolina University
Kathy Richardson, Mathematical Perspectives
Norma Sermon-Boyd, Jones County Schools
Judith T. Sowder, San Diego State University
William Spooner, North Carolina Department of Public Instruction
Diane M. Spresser, National Science Foundation
Chuck Thompson, University of Louisville

Clara Tolbert, Philadelphia Public Schools
Jean Vanski, National Science Foundation
Norman L. Webb, University of Wisconsin
Jan Williamson, North Carolina Department of Public Instruction
Lisa S. Williamson, Pinnacle, NC, Public Schools
Brenda Wojnowski, University of North Carolina Mathematics and
    Science Education Network
Judith S. Zawojewski, University of Pittsburgh

Bibliography of Books:
Classroom Assessment in Mathematics

Barnett, C., Goldenstein, D., & Jackson, B. (Eds.). (1994). *Fractions, decimals, ratios, and percents: Hard to teach and hard to learn? Facilitator's discussion guide.* Portsmouth, NH: Heinemann.

Barnett, C., Goldenstein, D., & Jackson, B. (Eds.). (1994). *Fractions, decimals, ratios, and percents: Hard to teach and hard to learn?* Portsmouth, NH: Heinemann.

Berenson, S., & Carter, G. (1995). *Alternative assessments: Practical applications for mathematics and science teachers.* Raleigh, NC: Center for Research in Mathematics and Science Education.

Birenbaum, M., & Dochy, F. J. R. C. (Eds.). (1996). *Alternatives in assessment of achievements, learning processes and prior knowledge.* Boston, MA: Kluwer Academic Publishers.

he examinations: An international comparison of science and mathematics examinations for college-bound students. Boston, MA: Kluwer Academic Publishers.

Carpenter, T. P., Corbitt, M. K., Kepner, H. S., Jr., Lindquist, M. M., & Reys, R. E. (1981). *Results from the Second Mathematics Assessment of the National Assessment of Educational Progress.* Reston, VA: National Council of Teachers of Mathematics.

Carpenter, T. P., Coburn, T. G., Reys, R. E., & Wilson, J. W. (1978). *Results from the First Mathematics Assessment of the National Assessment of Educational Progress.* Reston, VA: National Council of Teachers of Mathematics.

Chuska, K. R. (1995). *Improving classroom questions: A teacher's guide to increasing student motivation, participation, and higher-level thinking.* Bloomington, IN: Phi Delta Kappa Educational Foundation.

Danielson, C. (1996). *A collection of performance tasks and rubrics: Upper elementary school mathematics.* Larchmont, NY: Eye on Education.

Danielson, C. (1997). *A collection of performance tasks and rubrics: Middle school mathematics.* Larchmont, NY: Eye on Education.

Danielson, C. (in press). *A collection of performance tasks and rubrics: Primary school mathematics.* Larchmont, NY: Eye on Education.

Danielson, C., & Marquez, E. (in press). *A collection of performance tasks and rubrics: High school mathematics.* Larchmont, NY: Eye on Education.

Darling-Hammond, L., Ancess, J., & Falk, B. (1995). *Authentic assessment in action: Studies of schools and students at work.* New York, NY: Teachers College Press.

Fennema, E., & Nelson, B. S. (Eds.). (1997). *Mathematics teachers in transition.* Hillsdale, NJ: Lawrence Erlbaum Associates.

Friel, S. N., & Bright, G. W. (Eds.). (1997). *Reflecting on our work: NSF teacher enhancement in K-6 mathematics.* Lanham, MD: University Press of America.

Gal, I., & Garfield, J. (1997). *The assessment challenge in statistics education.* Amsterdam, The Netherlands: International Statistical Institute Press.

Gifford, B. R., & O'Connor, M. C. (Eds.). (1992). *Changing assessments: Alternative views of aptitude achievement and instruction.* Boston, MA: Kluwer.

Hiebert, J. (1988). *Conceptual and procedural knowledge: The case of mathematics.* Hillsdale, NJ: Lawrence Erlbaum Associates.

Kellaghan, T., Madaus, G. F., & Raczek, A. (1996). *The use of external examinations to improve student motivation.* Washington, DC: American Educational Research Association.

Kenney, P. A., & Silver, E. A. (Eds.). (1997). *Results from the Sixth Mathematics Assessment of the National Assessment of Educational Progress.* Reston, VA: National Council of Teachers of Mathematics.

Kuhs, T. M. (1997). *Measure for measure: Using portfolios in K-8 mathematics.* Portsmouth, NH: Heinemann.

Kulm, G. (Ed.). (1990). *Assessing higher order thinking in mathematics.* Washington, DC: American Association for the Advancement of Science.

Kulm, G. (1994). *Mathematics assessment: What works in the classroom.* San Francisco, CA: Jossey-Bass.

Kulm, G., & Malcom, S. M. (Eds.). (1991). *Science assessment in the service of reform.* Washington, DC: American Association for the Advancement of Science.

Lambdin, D. V., Kehle, P. E., & Preston, R. V. (1996). *Emphasis on assessment: Readings from NCTM's school -based journals.* Reston, VA: National Council of Teachers of Mathematics.

Leder, G. (Ed.). (1992). *Assessment and learning of mathematics*. Victoria, Australia: Australian Council for Educational Research, Ltd.

Lesh, R., & Lamon, S. J. (Eds.). (1992). *Assessment of authentic performance in school mathematics*. Washington, DC: American Association for the Advancement of Science.

Lindquist, M. M. (Ed.). (1989). *Results from the fourth mathematics assessment of the National Assessment of Educational Progress*. Reston, VA: National Council of Teachers of Mathematics.

Madaus, G. F., West, M. M., Harmon, M. C., Lomax, R. G., & Viator, K. T. (1992). *The influence of testing on teaching math and science in grades 4-12*. Chestnut Hill, MA: Center for Testing, Evaluation and Educational Policy, Boston College.

Mathematical Sciences Education Board. (1993). *Measuring what counts: A conceptual guide for mathematics assessment*. Washington, DC: Author.

McMillan, J. H. (1996). *Classroom assessment: Principles and practice for effective instruction*. Boston, MA: Allyn and Bacon.

McTighe, J., & Ferrara, S. (1994). *Assessing learning in the classroom*. Washington, DC: National Education Association.

Mokros, J. (1996). *Beyond facts and flashcards: Exploring math with your kids*. Portsmouth, NH: Heinemann.

Moon, J. (1997). *Developing judgment: Assessing children's work in mathematics*. Portsmouth, NH: Heinemann.

Moon, J., & Schulman, L. (1995). *Finding the connections: Linking assessment, instruction, and curriculum in elementary mathematics*. Portsmouth, NH: Heinemann.

National Academy of Science. (1989). *Everybody counts*. Washington, DC: Author.

National Council of Teachers of Mathematics. (1989). *Curriculum and evaluation standards for school mathematics*. Reston, VA: Author.

National Council of Teachers of Mathematics. (1991). *Professional standards for teaching mathematics*. Reston, VA: Author.

National Council of Teachers of Mathematics. (1995). *Assessment standards for school mathematics*. Reston, VA: Author.

National Research Center on Assessment, Evaluation, and Testing. (1990). *Standards for teacher competence in educational assessment of students*. Berkeley, CA: University of California.

Nelson, B. S. (Ed.). (1995). *Inquiry and the development of teaching: Issues in the transformation of mathematics teaching*. Newton, MA: Education Development Center.

New Standards. (1997). *1996 New Standards reference examination technical summary: A report of the New Standards*

Technical Studies Unit. Pittsburgh, PA: Learning Research & Development Center, University of Pittsburgh.

Parker, R. (1993). *Mathematical power: Lessons from a classroom.* Portsmouth, NH: Heinemann.

Reese, C. M., Miller, K. E., Mazzeo, J., & Dossey, J. A. (1997). *NAEP 1996 mathematics report card for the nation and the states: Findings from the National Assessment of Educational Progress.* Washington, DC: National Center for Education Statistics.

Richardson, K. (in press). *How do we know they're learning? Assessing math concepts.* Bellingham, WA: Lummi Bay Press.

Romberg, T. A. (Ed.). (1992). *Mathematics assessment and evaluation: Imperatives for mathematics educators.* Albany, NY: State University of New York Press.

Romberg, T. A. (Ed.). (1995). *Reform in school mathematics and authentic assessment.* Albany, NY: State University of New York Press.

Schifter, D. (Ed.). (1996). *What's happening in math class? Volume 1: Envisioning new practices through teacher narratives.* New York, NY: Teachers College Press.

Schifter, D. (Ed.). (1996). *What's happening in math class? Volume 2: Reconstructing professional identities.* New York, NY: Teachers College Press.

Schmidt, W. H., McKnight, C. C., & Raizen, S. A. (1996). *Splintered Vision: An Investigation of U.S. Science and Mathematics Education, Executive Summary.* East Lansing, MI: Michigan State University, U.S. National Research Center for the Third International Mathematics and Science Study.

School Curriculum and Assessment Authority. (1997, March). *The teaching and assessment of number at key stages 1-3.* London, England: Author.

Stenmark, J. (1991). *Mathematics assessment: Myths, models, good questions, and practical suggestions.* Reston, VA: National Council of Teachers of Mathematics.

Treagust, D. F., Duit, R., & Fraser B. J. (Eds.). (1996). *Improving teaching and learning in science and mathematics.* New York, NY: Teachers College Press.

U.S. Department of Education, National Center for Education Statistics. (1996). *Pursuing Excellence,* NCES 97-198. Washington, DC: U.S. Government Printing Office.

Van den Heuvel-Panhuizen, M. (1996). *Assessment and realistic mathematics education.* Utrecht, The Netherlands: Freudenthal Institute.

Webb, N. L. (1997). *Criteria for alignment of expectations and assessment in mathematics and science education* (Research Monograph No. 6). Madison: National Institute for Science Education, Wisconsin Center for Education Research, University of Wisconsin.

Webb, N. L., & Coxford, A. F. (Eds.). (1993). *Assessment in the mathematics classroom: 1993 yearbook.* Reston, VA: National Council of Teachers of Mathematics

Wiggins, G. P. (1993). *Assessing student performance: Exploring the purpose and limits of testing.* San Francisco, CA: Jossey-Bass.